Caribbean Primary

Mathematics

6th edition

Level 4

Contributors

Jonella Giffard

Hyacinth Dorleon

Melka Daniel

Martiniana Smith

Troy Nestor

Lydon Richardson

Rachel Mason

Eugenia Charles

Sharon Henry-Phillip

Glenroy Phillip

Jeffrey Blaize

Clyde Fitzpatrick

Reynold Francis

Wilma Alexander

Rodney Julien

C. Ellsworth Diamond

Shara Quinn

HODDER
EDUCATION
AN HACHETTE UK COMPANY

Contents

How to use this book

This Student's Book meets the objectives of the OECS and regional curricula for Level 4. The Student's Book provides both learning notes marked as [Explain] and a wide range of activities to help and encourage students to meet the learning outcomes for the level.

The content is arranged in topics, which correspond to the strands in the syllabus. For example, Topics 2, 5 and 9 relate to the strand of Number sense, while Topics 4, 12 and 16 relate to the various aspects that need to be covered in the strand, Geometry (Shape and space).

Topic 1, **Getting ready**, is a revision topic that allows you to do a baseline assessment of key skills and concepts covered in the previous level. You may need to revise some of these concepts if students are uncertain or struggle with them.

Each topic begins with an opening spread with the following features:

Notes for teachers about key mathematical skills that need to be developed.

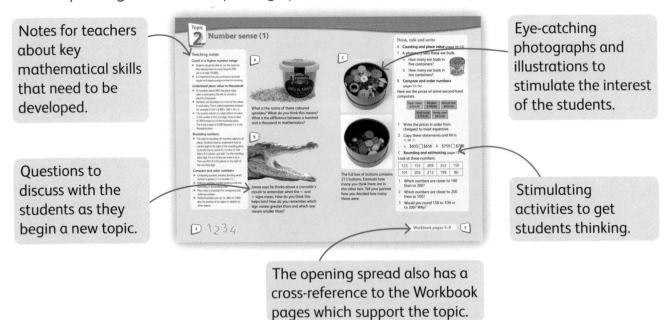

Eye-catching photographs and illustrations to stimulate the interest of the students.

Questions to discuss with the students as they begin a new topic.

Stimulating activities to get students thinking.

The opening spread also has a cross-reference to the Workbook pages which support the topic.

The questions and photographs relate to the specific units of each topic. You can let the students do all the activities (A, B and C, for example) as you start a topic, or you can do just the activities that relate to the unit you are about to start.

Topics 2 to 16 are sub-divided into units which deal with different skills that need to be developed to meet the learning outcomes. For example, Topic 2, **Number sense (1)** is divided into three units: **A** Counting and place value, **B** Compare and order numbers and **C** Rounding and estimating.

Each unit is structured in a similar way.

A student-friendly list of learning objectives.

Key word list. The words are bold and blue in the text.

Teaching text which explains concepts and provides examples.

Graded activities for consolidation.

Problem-solving activities.

Important symbols are highlighted.

A review activity.

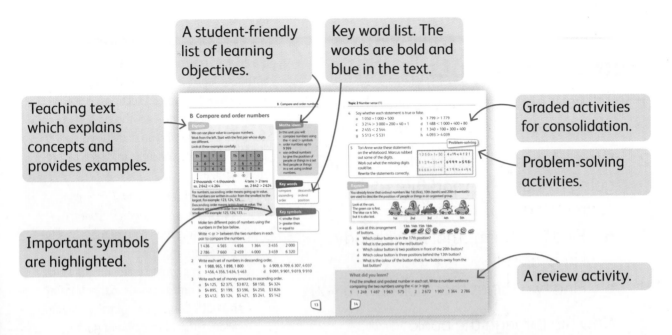

As you work through the units you will find a range of different types of activities and tasks, including practical investigations, problem-solving strategies, projects and challenge questions. These features are clearly marked in the book so you know what you are dealing with.

The topics end with a review page that provides:

* An interactive summary called *Key ideas and concepts* to help students consolidate and reflect on what they've learnt.

* *Think, talk, write* … activities which encourage students to share ideas, clarify their thinking and write in their maths journals.

* *Quick check* revision questions that cover all the units.

There are three tests provided in the Student's Book to allow for ongoing assessment and to prepare students for formal testing at all levels. Test 1 covers work from Topics 1–6, Test 2 covers work from Topics 1–12 and Test 3 covers work from the entire book.

Getting ready

Number concepts

1 Tell your partner what you liked best about mathematics last year.

2 Work in pairs to find out what you are going to learn this year. Turn to the contents page of this book.

   a Find two things that you already know about.

   b Find two things that look new to you.

3 Now flip through the book.

   a Find three things that look interesting.

   b Show your partner the things you have found. Tell each other why these things interest you.

4 What are your goals for mathematics this year? Are there things you would like to improve? How will you try to improve? Write down your goals and how you plan to meet them in your journal.

5 Here is a set of numbers. Say each number out loud before you start.

345	198	989	$\frac{5}{9}$	206	$\frac{3}{5}$

   a Write the whole numbers in descending order.

   b Which number is a fraction with a denominator of 5?

   c Write two fractions that are equivalent to $\frac{2}{3}$.

   d Which number has 9 in the tens place?

Teaching notes

In this topic we revisit some of the key skills taught in Level 3.

You can use these pages at the beginning of the year, as part of your baseline assessment of the students' knowledge and skills, or you can use the relevant sections as you begin each new topic with the class. If your students struggle with an activity, you will need to revise the concepts and skills before moving on.

∗ Can students read and write numbers to 999 in figures and words? If not, use flashcards and matching games to revise and reinforce the concepts.

∗ Can students count confidently forwards, backwards and in groups in the range of 0 to 1 000? Continue to practice counting regularly: start from different numbers, and count both forwards and backwards. Display a 101–200 chart in the classroom to remind students of counting patterns in the hundreds.

∗ Do students understand place value in two- and three-digit numbers? If not, continue to work with place value tables to break down numbers using place value. Let students model numbers using blocks, if necessary.

∗ Are students able to classify numbers as odd and even and explain their choices? If not, you may need to reteach this skill using modelling to show that even numbers are all divisible by 2 and asking students to write their own rules for classifying numbers by looking at the digits in the units place.

∗ Can students identify and compare fractions and use the terms numerator and denominator? If not, revise the basics using pictorial and concrete representations. Also play games in which students find a fraction of a whole or group.

∗ Are students confidently using the < and > symbols to compare whole numbers and fractions? Remind them what the symbols represent, if necessary.

Workbook pages 1–4

    e Which two whole numbers are odd numbers? How do you know this?

    f Write a new number that lies between 500 and 600. It should be an even number with a 7 in the tens place and a number < 6 in the ones place.

    g Write a number that is 100 more than 345.

    h What number is 10 less than 206? Write it in words.

6 Count from 180 to 220 in fives. Write the numbers you count. Circle the even numbers.

7 Rewrite these statements. Fill in $<$, $=$ or $>$ to make each statement true.

    a 132 ☐ 100 + 30 + 2   **b** 342 ☐ 432   **c** 543 ☐ 534

8 Answer these questions about the number 649.

    a Which digit is in the tens place?

    b What is the value of the 6?

    c What is the difference between the values of the 4 and the 9?

    d If 649 is the last number in a pattern with the rule 'subtract 2', what are the two numbers before it?

    e If 649 is the first number in the pattern 'add 3', what are the next three numbers in the pattern?

    f Round 649 off to the nearest 10.

9 Copy and complete each number pattern.

    a 4, 8, 12, ___, ___, 24, 28   **b** 35, 32, 29, ___, ___, 20, 17

    c 2, 4, 8, 14, ___, ___, 44, 58   **d** 11, 13, ___, 17, 23, ___, 25

10 Two thirds of each shape is shaded blue. Use the diagrams to help you answer the questions below.

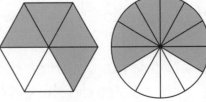

    a $\frac{2}{3} = \frac{4}{\square}$ **b** $\frac{2}{3} = \frac{\square}{12}$

    c Which fraction is greater: $\frac{2}{3}$ or $\frac{7}{9}$?

11 Martin finished 12th in a race.

    a How many people finished before him?

    b Four people finished after Martin. In what position was the last person?

12 Nadia has six sheets with 10 stickers and one sheet with only 7 stickers on it.

    a How many stickers does she have altogether?

    b How many more does she need to have 100?

Computation

1   Write only the answers in your exercise book.

   a $7 + 8$ **b** $25 + 10$ **c** 9×6

   d $40 \div 2$ **e** double 12   **f** 5×5

   g $19 + 20$ **h** $25 - 9$

2   Do these calculations. Show how you work each one out.

   a $234 + 567$ **b** $299 + 403$

   c $299 - 145$ **d** $416 - 224$

   e 23×4 **f** 87×5

   g 23×10 **h** 69×10

3   Say whether each number sentence is true or false. If it is false, write it correctly.

   a $345 + 1 > 356$ **b** $98 \times 1 = 981$

   c $302 + 22 < 322$ **d** $20 \times 8 < 8 \times 20$

4   Round off each number and estimate the answer.

   a $341 + 408$ **b** $699 + 129$

   c $987 - 235$ **d** $872 - 458$

Problem-solving

5   At the Agape Chocolate Factory you can buy handmade chocolates. One box holds 12 chocolates.

   a Sal ate half a box. How many chocolates were left?

   b Nicole bought three boxes. How many chocolates did she get?

   c Nadia and Nikita bought three boxes to share equally. How many chocolates did they each get?

   d Simon got two boxes as a present. He ate $\frac{1}{4}$ of the chocolates. How many did he eat?

Teaching notes

Check:

✳ Can students use and understand the vocabulary of addition, subtraction, multiplication and division? Use flashcards with the words and let students explain their meanings. Then play matching games in which students match words, meanings and operations.

✳ Do students know the basic addition and subtraction facts? Let them take turns to test each other. Aim for a three-second recall.

✳ Can students use different strategies to recall multiplication and related division facts? Let students explain their strategies to each other and use them to revise and memorise the basic facts.

✳ Can students do written calculations with up to three-digit numbers? Let them check each other's work and observe them as they do calculations so that you can correct any misconceptions or mistakes.

✳ Can students add fractions with the same denominators? Continue to use concrete models and diagrams to demonstrate and model this.

✳ Can they make sense of problems and use appropriate strategies to model and solve them? Continue to talk through the problem-solving process as you model different approaches with the class. Revise the bar diagram methods of drawing to model problems if needed.

Shapes and patterns

1   a   Write the name of each shape and say whether it is a 2-D or 3-D shape.

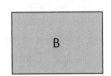

   b   How many sides does a square have? What is special about squares?

   c   How many faces does a cuboid have? What shape are cuboids?

2   a   Say whether each figure is an open or closed shape.

   b   Which shape has two long sides and two short sides? What is its name?

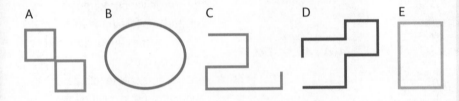

3   Look at this picture of a house. What shapes can you see?

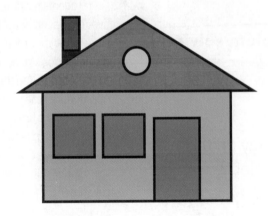

Teaching notes

Check:

* Can students identify, name and sort the basic 2-D shapes? If not, revise the names using cut-out shapes and patterns in the environment.

* Can students identify, name and sort the basic 3-D shapes? If not, use solid objects to revise and reinforce the names.

* Can students group 2-D shapes using shape properties? Remind students to count sides and compare the lengths of sides and the sizes of angles. Ask questions like: How many shapes can you see with four sides? Which of these have all their sides the same length? Which shapes have only three sides?

* Can students recognise and name line segments and right angles? Revise the terms as necessary and let the students make a folded 'corner' to measure angles to see if they are greater or smaller than a right angle. Make sure students know how to measure and draw line segments of a given length.

* Are students able to say whether a shape is symmetrical or not? Can they identify and draw lines of symmetry on diagrams? Encourage the students to trace and fold shapes to see if they are symmetrical, if they struggle to do this visually. Use small mirrors to check for symmetry to reinforce concepts in a fun way.

5

Measurement and statistics

1   Look at the clock.

   a   What time does it show?

   b   What will the time be
       25 minutes later than this?

2   What measuring instrument and which unit
   of measurement would you use to measure
   each item?

   a   your own weight

   b   your height

   c   the mass of a cup of sugar

3   What is the perimeter of each shape?

   a   **b**

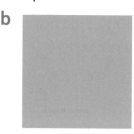

4   Use a calendar.

   a   How many weekdays are there this month?

   b   What day and date is the last day of
       this month?

   c   How many weeks is 21 days?

Problem-solving

5   Six children received an
   equal share of $48.

   a   How much did they each get?

   b   One girl spent half of her money on the
       way home. What did she have left?

   c   One boy spent $2 on snacks and gave 50¢
       to charity. How much did he have left?

6   Mrs Smythe gets on a bus at twenty to
   twelve. If she is on the bus for an hour and a
   quarter, at what time does she get off?

Teaching notes
Check:

* Are students able to use
   appropriate measuring
   instruments and standard units
   to measure length, mass and
   capacity? Continued use of
   measuring instruments and
   units in practical situations
   is the most effective way to
   improve measuring skills.

* Do students understand the
   concept of perimeter? Can they
   measure and calculate perimeter?
   Make sure students know that
   they need to add the lengths of
   all sides of a shape and that if
   these are given on a diagram,
   they should not measure.

* Can students state and write
   the date in various formats?
   Let students take turns each
   day to say the current date and
   write it on the board.

* Are students able to talk
   about time using appropriate
   vocabulary? Ask questions like:
   When do we ...? What did you do
   yesterday? What day will it be
   tomorrow? And so on.

* Can students tell time up to
   five-minute intervals? Display a
   clock in the classroom or have
   a smaller clock for each group
   to use. Stop during the day at
   appropriate times to let them
   work out what time it is.

* Are students able to work with
   different units of time and state
   the relationship between these?
   Let students make up and solve
   simple problems involving units
   of time. For example, how many
   weeks in two years? Use a clock
   and calendar regularly to revise
   and reinforce the relationships
   between units.

7 How can you collect information about what TV programmes people like best?

8 How can you record this information?

9 Can you predict what the results of a survey will be?

10 Why is it useful to collect data? Give an example.

11 Look at this graph. Explain why it is not very useful.

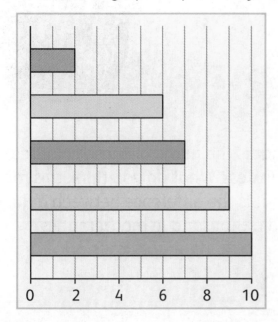

12 Keisha and Torianne collected data about the number of children in different levels who wear glasses. They made this table to show their results.

Number of children who wear glasses	
Level 3	₩₩ ₩₩ ₩₩ //
Level 4	₩₩ ₩₩ ₩₩ ////
Level 5	₩₩ ₩₩ ₩₩
Level 6	₩₩ ₩₩ ₩₩ ₩₩ ///

  a What kind of table is this?

  b Which level has 19 children who wear glasses?

  c How many children wear glasses in Level 6?

  d How many children wear glasses in all four levels?

  e Draw a bar graph to show this data.

✳ Can students describe the money that is in use? Use real coins and notes or pictures of these to reinforce the concepts. Play games involving counting money, paying amounts and making up amounts in different ways so students become familiar with the money in use and its value.

✳ Are students able to do simple calculations involving money? Set up a class shop and play trading games in which students tender amounts for a set of items and give and get change.

✳ Can students collect simple sets of data by observation and by asking questions? Help students who struggle by providing frames and tables to use to collect and organise data.

✳ Are students able to use tally charts, tables and graphs? Let students complete partly filled-in tables, pictographs and bar graphs to revise and reinforce the skills and concepts.

✳ Can students look at a graph and find the answers to questions? Model how to find answers on a graph with the class and teach students to read titles, labels and keys carefully before they answer questions about the graph.

Teaching notes

Count in a higher number range

* Students should be able to use the patterns they already know to count beyond 999 (up to at least 10 000).
* It is important that you continue to provide regular and ongoing opportunities for counting.

Understand place value to thousands

* For numbers above 999, the place value table is extended to the left to include a place for thousands.
* Numbers can be written as a sum of the values in each place. This is called expanded notation. For example: 6 345 = 6 000 + 300 + 40 + 5.
* The position (place) of a digit affects its value. In the number 6 345, the digit 3 has a value of 300 because it is in the hundreds place. The 6 has a value of 6 000 because it is in the thousands place.

Rounding numbers

* The rules for rounding off numbers apply to all places. Students need to understand how to use the digit to the right of the rounding place to decide how to round the number. If that digit is 5 or greater, you add 1 to the rounding place digit, if it is 4 or less you leave it as is. Then you fill in 0 in the places to the right of the rounding digit.

Compare and order numbers

* Comparing numbers involves deciding which number is greater (>) or smaller (<).
* Ordering numbers involves putting a set into ascending or descending order.
* Place value is important for comparing and ordering numbers.
* Ordinal numbers such as 1st, 28th or 100th give the position of an object in relation to other objects.

A

What is the name of these coloured sprinkles? What do you think this means? What is the difference between a hundred and a thousand in mathematics?

B

James says he thinks about a crocodile's mouth to remember what the < and > signs mean. How do you think this helps him? How do you remember which sign means greater than and which one means smaller than?

1234

C

The full box of buttons contains 213 buttons. Estimate how many you think there are in the other box. Tell your partner how you decided how many there were.

Think, talk and write

A **Counting and place value** *(pages 10–12)*

1 A pharmacy sells these ear buds.

a How many ear buds in five containers?

b How many ear buds in ten containers?

B **Compare and order numbers** *(pages 13–14)*

Here are the prices of some second-hand computers.

Super cheap $255.00	Bargain $700.00	Almost new $605.00

Much loved $759.00	Barely used $650.00

1 Write the prices in order from the cheapest to the most expensive.

2 Copy these statements and fill in < or >.

a $605 ☐ $650   b $759 ☐ $700

C **Rounding and estimating** *(pages 15–16)*

Look at these numbers:

123	155	209	245	150
101	203	212	199	80

1 Which numbers are closer to 100 than to 200?

2 Which numbers are closer to 200 than to 100?

3 Would you round 150 to 100 or to 200? Why?

A Counting and place value

Last year you worked with 3-digit numbers and counted up to one thousand. Now you are going to learn about 4-digit numbers and count up to ten thousand.

The number 328 has three **digits**. We say three hundred and twenty-eight.

We can show three-digit numbers on a **place value** table like this:

hundreds	tens	ones
3	2	8

We use **expanded notation** to write a number as the sum of the **values** in each place.

328 can be written in expanded form like this:
328 = 300 + 20 + 8

The place of each digit tells us what value the digit has in the number. In the number 328, the 3 has a value of 3 hundreds or 300, because it is in the hundreds place. The 2 has a value of 2 tens or 20, because it is in the tens place. The 8 has a value of 8 ones or 8, because it is in the ones place.

You can only write digits up to 9 in each place.
999 is the highest three-digit number.
999 + 1 makes a **thousand**. We write 1 000.

1 000 is a 4-digit number. It has four digits. Each digit has its own place value.

The place value table for numbers in the thousands looks like this:

thousands	hundreds	tens	ones
3	1	4	5

3 145 is three thousand one hundred and forty-five.
The 3 is in the thousands place, so it has a value of 3 thousands or 3 000.

We can write this in expanded form like this: 3 145 = 3 000 + 100 + 40 + 5.

The number 2 104 has no tens, so we write a 0 in the tens place as a place holder. When you write this number in expanded notation, it is:
2 104 = 2 000 + 100 + 4

You do not need to write the 0 in the sum, but you do need to write it in the number because 2 104 is a different number from 214.

In this unit you will:
* extend the place value table to include thousands, hundreds, tens and units
* state the place value and total value of any digit in numbers up to 10 000
* use expanded notation to write numbers
* count forwards and backwards from any number between 0 and 10 000.

Key words

digits
place value
expanded notation
values
thousand

1   Read each number.

a	365	b	205	c	879	d	965
e	100	f	2400	g	3250	h	9000
i	2456	j	4050	k	3002	l	4098

2   Write these numbers using numerals.

a   two thousand three hundred

b   four thousand five hundred and twenty-seven

c   eight thousand three hundred and forty-nine

d   nine thousand nine hundred and ninety-nine

e   five thousand and fifty   f   five thousand and five

3   What is the value of the 2 in each of these numbers?

a	235	b	325	c	532	d	523
e	2614	f	3215	g	5124	h	4132

4   Follow these instructions. Write the first five numbers that you count.

a   Start at 3500. Count forwards in ones.

b   Start at 4250. Count backwards in ones.

c   Start at 900. Count forwards in hundreds.

d   Start at 2400. Count backwards in hundreds.

e   Start at 3459. Count forwards in tens.

f   Start at 2564. Count backwards in tens.

5   Read the numbers in the blocks.

3254	6024	4581	5124

Write the number that has:

a   4 ones and no hundreds   b   1 one and 5 hundreds

c   4 thousands and 5 hundreds   d   2 hundreds and 5 tens.

Problem-solving

6   Use the digits 1, 3, 5 and 7 once only.

a   What is the biggest number you can make?

b   What is the smallest number you can make?

c   How many different 4-digit numbers can you make using only these four digits?

7   Write each of these numbers in expanded notation.
   a   24   b   98   c   129   d   450
   e   5 465   f   2 309   g   3 120   h   4 060
   i   8 765   j   9 990   k   9 099   l   5 498

8   Write these amounts as numbers.
   a   3 000 + 300 + 50 + 2   b   5 000 + 400 + 20 + 5
   c   6 000 + 40 + 300   d   400 + 30 + 2 + 2 000
   e   9 + 5 000 + 40   f   9 + 500 + 2 000 + 30

9   Write the missing value in each expanded notation.
   a   2 345 = 2 000 + 300 + ☐ + 5   b   4 234 = ☐ + 200 + 30 + 4
   c   9 876 = 9 000 + ☐ + 70 + 6   d   4 027 = 4 000 + ☐ + 7
   e   3 006 = 3 000 + ☐   f   4 060 = 4 000 + ☐

10 Read each number and then write it in words.
   a   876   b   6 436   c   2 090   d   4 650
   e   5 439   f   4 999   g   3 800   h   1 909

Challenge

11 Benita cut these shapes out of number charts and rubbed out
    some of the numbers.
   a   Try to work out
       which numbers
       she rubbed out.
   b   How did you use
       number patterns
       to help you?

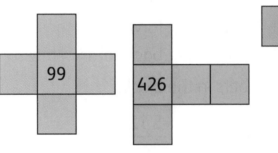

What did you learn?

Write the number shown on each of these place value tables in figures, in
expanded notation and in words.

1
Th	H	T	O
••	•	••	•

2
Th	H	T	O
••	•		••••

3
Th	H	T	O
••••	••		••••••

B Compare and order numbers

Explain

We can use place value to **compare** numbers.
Work from the left. Start with the first pair whose digits are different.
Look at these examples carefully.

Th	H	T	O
2	6	4	2
4	2	6	4

Th	H	T	O
2	6	4	2
2	6	2	4

2 thousands < 4 thousands
so, 2 642 < 4 264

4 tens > 2 tens
so, 2 642 > 2 624

For numbers, **ascending** order means going up in value.
The numbers are written in **order** from the smallest to the largest. For example: 123, 124, 125, …
Descending order means going down in value. The numbers are written in order from the largest to the smallest. For example: 125, 124, 123, …

Maths ideas

In this unit you will:
* compare numbers using the < and > symbols
* order numbers up to 9 999
* use ordinal numbers to give the position of people or things in a set
* find people or things in a set using ordinal numbers.

Key words

compare	descending
ascending	ordinal
order	position

Key symbols

< smaller than
> greater than
= equal to

1. Make ten different pairs of numbers using the numbers in the box below.

   Write < or > between the two numbers in each pair to compare the numbers.

1 436	4 565	4 656	1 364	3 455	2 000
2 786	7 660	2 459	4 000	3 459	6 320

2. Write each set of numbers in descending order.
   a 1 988, 965, 1 898, 1 800
   b 4 909, 6 709, 6 307, 4 037
   c 3 456, 4 356, 5 634, 5 463
   d 9 091, 9 901, 9 019, 9 910

3. Write each set of money amounts in ascending order.
   a $4 125, $2 375, $3 872, $8 150, $4 324
   b $4 895, $1 199, $3 596, $4 250, $3 826
   c $5 412, $5 124, $5 421, $5 241, $5 142

13

4   Say whether each statement is true or false.

   a $1\,050 = 1\,000 + 500$ b $1\,799 > 1\,779$

   c $3\,214 > 3\,000 + 200 + 40 + 1$ d $1\,488 < 1\,000 + 400 + 80$

   e $2\,455 < 2\,544$ f $1\,340 = 100 + 300 + 400$

   g $5\,513 < 5\,531$ h $4\,093 > 4\,039$

Problem-solving

5   Tori-Anne wrote these statements on the whiteboard. Marcus rubbed out some of the digits.

Work out what the missing digits could be.

Rewrite the statements correctly.

$1\,2\,5\,0 > 1\,\blacksquare\,50$	$4\,\blacksquare\,19 < 4\,1\,2\,1$
$5\,1\,2\,9 < 51\,\blacksquare\,9$	$6\,9\,9\,9 > 5\,9\,8\,\blacksquare$
$4\,6\,6\,6 > 4\,\blacksquare\,\blacksquare\,6$	$4\,1\,9\,9 > 4\,\blacksquare\,9\,9$

Explain

You already know that **ordinal** numbers like 1st (first), 10th (tenth) and 20th (twentieth) are used to describe the **position** of people or things in an organised group.

Look at the cars.
The green car is first.
The blue car is 5th, but it is also last.

   1st   2nd   3rd   4th   5th

6   Look at this arrangement of buttons.

13th 14th 15th 16th

   a   Which colour button is in the 17th position?

   b   What is the position of the red button?

   c   Which colour button is two positions in front of the 20th button?

   d   Which colour button is three positions behind the 13th button?

   e   What is the colour of the button that is five buttons away from the last button?

What did you learn?

Find the smallest and greatest number in each set. Write a number sentence comparing the two numbers using the $<$ or $>$ sign.

1   1249   1497   1963   575   2   2672   1907   1364   2786

C Rounding and estimating

Explain

Do you remember how to use **place value** to round numbers to a given place?

Find the **digit** in the rounding place.
* Look at the digit to the right of this place.
* If the digit to the right is 0, 1, 2, 3 or 4, leave the digit in the rounding place as it is.
* If the digit to the right is 5, 6, 7, 8 or 9, add 1 to the digit in the rounding place.
* Change all the digits to the right of the rounding place to 0.

Example

Round 8 763 to the **nearest** hundred.

This is the hundreds place.

8 7 6 3

8 8 0 0

Write 0s in all the places to the right.

Maths ideas

In this unit you will:
* revise the rules for rounding numbers
* round numbers to the nearest ten and hundred
* use rounded numbers to estimate answers.

Key words

place value   rounded

digit   estimated

nearest

1 Round each red number to the nearest ten and the nearest hundred.

a
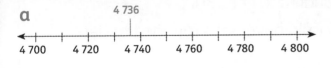

b
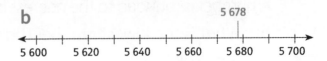

c
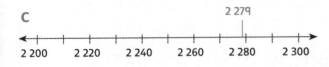

d
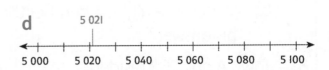

2 Round each number to the nearest ten.

a 1089	b 76	c 2603
d 1721	e 9432	f 719
g 9455	h 99	

3 Round each number to the nearest hundred.

a 1345	b 5570	c 3470
d 9876	e 2499	f 2654
g 2987	h 4108	

Think and talk

Two students **rounded** the number 6 192.

Sharon wrote 6 200.

Reynard wrote 6 190.

Their teacher said they were both correct, so why did they get different answers?

15

4   List all the numbers in the box that will give 1600 when they are rounded to the nearest hundred.

1509	1589	1549	1602	1650	1550	1702	1649

5   Say whether each statement is true or false.

a   8731 rounds to 8700   b   1298 rounds to 1300

c   1245 rounds to 1250   d   2810 rounds to 2800

e   1298 rounds to 1300   f   3999 rounds to 4000

6   Read what these children are saying. How do you know they are using **estimated** amounts?

About 10 000 people attended the jazz festival.

Hundreds of people danced in the street during the carnival.

I've got about two hundred dollars in a savings account.

I'll be home around 4 o'clock.

Problem-solving

7   A number is rounded to the nearest hundred to get 2300.

a   What is the lowest number that could have been rounded to get this answer?

b   What is the highest number that could have been rounded to get this answer?

8   Town A has 762 residents, Town B has 1254 residents and Town C has 930 residents. Round the numbers to the nearest 100 and estimate how many people live in the three towns altogether.

What did you learn?

1   Round each number to the nearest ten.

a   2345   b 9806   c   999   d   1344

2   Which of these numbers round to 1300?

1367	1394	1381	1365	1276	1228

Topic 2 Review

Key ideas and concepts

Read the statements. Fill in the missing words to summarise what you learnt in this topic.

1   A number in the ____ has four digits.

2   Each digit in a number has its own ____ value.

3   Writing a number like this: 2 000 + 400 + 40 + 3 = 2 443 is called ____.

4   The numbers 1234, 1235, 1236, 1237 are written in ____ order.

5   In 3 < 5, the symbol < means ____ than.

6   To round numbers to the nearest ten or hundred, you look at the digit to the ____ of the place you are rounding to.

7   If the digit is 5 or more you add to the rounding digit and write ____ as place holders in the other places.

8   An ____ number gives the position of an object in relation to other objects.

Think, talk, write ...

1   Read the clue. Write the number. Say the number in words.
   *   I am a four-digit number with more than eight thousands. I have the same number of tens and ones. I have three hundreds. The digit in the tens place has a value of 90.

2   Make up a clue of your own for a four-digit number. Swap with a partner and try to work out each other's numbers.

Quick check

1   Write each set of numbers in ascending order.
   a   2 354, 2 453, 2 345, 2 435
   b   3 453, 4 112, 2 356, 3 089

2   Say whether each statement is true or false.
   a   3 459 < 3 954   b   8 560 > 8 660   c   3 668 < 3 686

3   Start at 4 889. Count forwards. Write the 5th and 10th numbers you would count.

Problem-solving

4   Josh makes up a 4-digit password for his computer. He starts by making the smallest possible number he can make choosing from the digits 1, 2, 3, 4 and 5 without repeating any digits. Then he rounds the number to the nearest ten. What is his password?

Teaching notes

Recall of basic facts

* By now students should know all addition and subtraction facts to at least 20 and be able to recall these in three seconds or less.

* Try to start each lesson with a short mental activity so that students get a chance to use number facts and consolidate strategies (such as partitioning, compensating or bridging through multiples of 10). You can only expect students to develop fluency and confidence if you provide regular and ongoing practice.

Mental strategies

* Encouraging mathematical thinking and reasoning, involves making time to talk about how students worked out their answers. You can get students to model their solutions on the board and let them verbalise and explain what steps they took. Others will learn from these explanations, as exposure to different options and methods of working allows them to consider multiple strategies and choose the one that they find the easiest and most efficient.

* Modelling and talking through solutions also makes strategies visible to students who may not have fully understood them.

Pen-and-paper calculation

* Students will perform addition and subtraction that involves an exchange (regrouping) across place values. Treat this as a normal element of calculating rather than as something difficult.

* Allow students to model calculations using place value and number lines until they are comfortable using the more formal column methods.

* Remember that the word 'sum' only applies to addition, so avoid using it to talk about other operations.

Drawing bar models

* Continue to draw and use bar models to help students visualise problems and to see what they need to do to solve it. This is an important step in the problem-solving process and not something that you do as an add-on.

A

Maria uses 12 beads to make a bangle. In this bangle she used 6 red and 6 yellow beads. How many red beads would she need if she used only 4 yellow? What other numbers of red and yellow beads could she combine to get 12?

B

What is the total shown on the six dice? How would you group them into pairs to make it easier to add the numbers?

1234

C

Mrs Mackenzie paid $165 for one pair of spectacles and $149 for another pair. How much did the two pairs cost in total? What operation did you use to work out the answer?

D

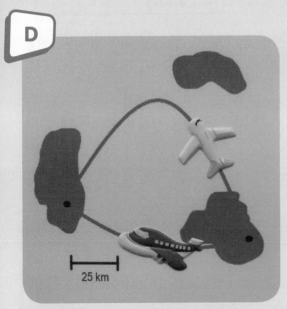

25 km

The red lines on the map show the routes two different aeroplanes flew. Which plane flew further? How much further did it fly? Tell your partner how you worked this out.

Think, talk and write

A   Number facts (pages 20–21)

How many ways can you find to make 13 by adding or subtracting pairs of numbers from 0 to 20? List all the ways you can find.

B   Mental strategies (pages 22–23)

1   Think of any number from 1 to 20. Double the number you chose. Then add 6. Halve the answer and subtract the number you started with. You should get 3.

2   Try this again with a few other numbers. Can you explain why you always get 3? Will this work if you start with 0?

C   Addition (pages 24–25)

1   Make up a sentence using each word in the box to show what it means.

> total   plus   add   sum
> altogether   combined   in all

2   A farmer planted 324 bean plants and 453 cabbage plants. How many plants is this altogether?

D   Subtraction (pages 26–28)

Read the problem. Work with a partner to decide how you would solve it.

Volunteers picked up litter on a beach. Group A picked up 863 pieces, group B picked up 758 pieces. How many more pieces did group A pick up than group B?

A Number facts

You already know all the **addition** and **subtraction** facts to 10 very well.

These **facts** can help you to add groups of numbers quickly by making tens.

Look at these examples carefully to see how to combine numbers to make tens.

$8 + 6 + 2 = \square$	$1 + 6 + 9 = \square$	$6 + 3 + 2 + 8 + 4 = \square$
$10 + 6 = 16$	$10 + 6 = 16$	$10 + 10 + 3 = 23$

This strategy can help you add larger numbers too.

30	10
$22 + 18 = \square$	$32 + 29 + 41 = \square$
10	90
$30 + 10 = 40$	$90 + 10 + 2 = 102$

Maths ideas

In this unit you will:
* revise the addition and subtraction facts you already know
* use facts you know to make calculations easier
* find and use patterns in addition and subtraction facts.

Key words

addition

subtraction

facts

fact family

---

1   Try to add these numbers by making tens mentally. Work as quickly as you can.

a $3 + 4 + 6$	b $4 + 5 + 5$	c $3 + 1 + 7$
d $4 + 2 + 8$	e $7 + 9 + 1$	f $7 + 3 + 9$
g $6 + 5 + 4$	h $5 + 8 + 5$	i $2 + 8 + 3$
j $2 + 6 + 8 + 4$	k $4 + 1 + 6 + 2$	l $7 + 1 + 3 + 9$
m $3 + 5 + 7 + 5$	n $9 + 1 + 9 + 3$	o $5 + 5 + 5 + 3$

Problem-solving

2   a   Match pairs of numbers from the box that give a combined total of 100.

91	27	32	19	45	88	42	63
23	48	58	39	77	37	81	12
55	9	73	40	52	61	60	68

   b   Tell your partner how you decided which numbers fit together and how you kept track of which ones you had already used.

Explain

You can break up a number to make tens to help you subtract.
Look at these examples to see how this works.

Example 1

$32 - 8 \longrightarrow 8 = 2 + 6$

Subtracting the 2 leaves $30 - 6$.

Now you can use subtraction facts to 10 to find the difference:

$30 - 6 = 24$

Example 2

$57 - 9 \longrightarrow 9 = 7 + 2$

Subtracting the 7 leaves $50 - 2$.

$50 - 2 = 48$

Challenge

If you add two numbers, you get 19. If you subtract the same two numbers, you get 5. What are the two numbers?

3   Try to do these subtractions mentally.

a	$23 - 8$	b	$43 - 7$	c	$63 - 9$	d	$53 - 4$
e	$31 - 9$	f	$41 - 8$	g	$51 - 6$	h	$91 - 4$
i	$72 - 8$	j	$42 - 9$	k	$32 - 7$	l	$92 - 6$
m	$35 - 8$	n	$45 - 7$	o	$85 - 9$	p	$95 - 6$

4   Do these subtractions.

a	$12 - 9$	b	$22 - 9$	c	$32 - 9$	d	$42 - 9$
e	$52 - 9$	f	$62 - 9$	g	$72 - 9$	h	$102 - 9$
i	$18 - 9$	j	$38 - 9$	k	$48 - 9$	l	$68 - 9$
m	$78 - 9$	n	$88 - 9$	o	$98 - 9$	p	$108 - 9$

5   Look at your answers in Question 4. What pattern do you notice? How can this help you subtract 9 from any number?

6   Complete each fact below. Then write the related facts for those three numbers to make a **fact family** for each.

a	$35 - 6$	b	$27 - 9$	c	$34 - 8$	d	$45 - 9$
e	$54 - 9$	f	$14 - 6$	g	$25 - 6$	h	$31 - 7$

What did you learn?

Calculate mentally. Write the answers only.

1   $6 + 4 + 1 + 8 + 1 + 5$

2   $5 + 6 + 4 + 5 + 3 + 7$

3   $11 - 7$

4   $21 - 7$

5   $31 - 7$

6   $51 - 7$

B Mental strategies

Mental strategies are the different ways we think and work things out when we calculate. Some mental strategies might include **jotting** things down.

Here are some useful strategies for adding and subtracting quickly.

Use the pattern of facts you already know

You know that 11 + 8 = 19, so use the same **pattern** to work out 110 + 80 = 190 and 1100 + 800 = 1900.

Break down numbers and use place value

Jottings can be useful when you **break down** numbers in this strategy.

40 + 50 = 90
45 + 53   98
5 + 3 = 8

60 − 20 = 40
68 − 27   41
8 − 7 = 1

Compensate so you can work with easier numbers

Compensation is useful when the units digit in one of the numbers is an 8 or a 9.

If you add more than you need, you must subtract it again at the end: 38 + 19 = 38 + 20 − 1 → 58 − 1 = 57

If you subtract more than you need, you must add it back again at the end: 48 − 29 = 48 − 30 + 1 → 18 + 1 = 19

Jump in chunks using jotting on a number line

This strategy uses jotting as a quick method of calculating. You can **jump** in whatever chunks you are comfortable using.

Example 1

167 + 86   Jump on from the greater number.

+40   +40   +3   +3

167   207   247   250   253

Example 2

456 − 239   Jump back from the greater number.

−3   −6   −30   −200

217   220   226   256   456

Maths ideas

In this unit you will:
* revise some of the mental strategies you used last year to add and subtract mentally
* learn some new strategies that you can use to calculate quickly
* practice using different strategies to see which numbers are suited to each method.

Key words

mental strategies
jotting
pattern
break down
compensation
jump
number lines

Hint

You can use any strategy that you think is best for each calculation, including **number lines**.

1   Use known facts and patterns to do these calculations. Try to write the
    answer only.
    a   50 + 50   b   70 + 30   c   90 + 10   d   110 + 50
    e   300 + 200   f   700 + 700   g   400 + 200   h   500 + 500
    i   600 – 200   j   100 – 90   k   2000 – 1500   l   700 – 300
    m   1200 – 800   n   1400 – 700   o   1500 – 1100   p   500 – 500

2   Use breaking down numbers and place value strategy to do these calculations.
    Try to work out the answers in your head. Remember that you can use jottings
    if you need to.
    a   47 + 43   b   81 + 34   c   87 + 26   d   43 + 88
    e   123 + 456   f   145 + 324   g   456 + 321   h   149 + 234
    i   89 – 34   j   67 – 36   k   94 – 65   l   87 – 38
    m   245 – 134   n   456 – 241   o   543 – 247   p   832 – 556

3   Use compensation to add or subtract.
    a   127 + 219   b   163 + 229   c   148 + 219   d   225 + 239
    e   376 + 219   f   254 + 229   g   129 + 325   h   499 + 235
    i   64 – 38   j   62 – 39   k   85 – 48   l   73 – 29
    m   134 – 129   n   245 – 218   o   283 – 118   p   345 – 219

4   Use the jump strategy to find the sum or difference. Draw a number line and
    show the jumps for each one.
    a   367 + 322   b   453 + 225   c   327 + 234   d   544 + 318
    e   346 + 235   f   765 + 328   g   663 + 438   h   488 + 97
    i   741 – 312   j   734 – 215   k   432 – 114   l   533 – 314
    m   645 – 326   n   832 – 199   o   500 – 245   p   825 – 498

What did you learn?

1   Calculate.
    a   23 + 19   b   132 – 29   c   312 + 578   d   764 – 213
    e   400 + 200   f   1800 – 900   g   320 – 145   h   582 – 349

2   Tell you partner what strategy you used to find each answer. Explain why you
    chose that method of working.

C Addition

The methods of **addition** that you learnt last year can be used to **add** numbers in the thousands.

Work through the examples carefully.

Use chunks on a number line: 2 045 + 1 832

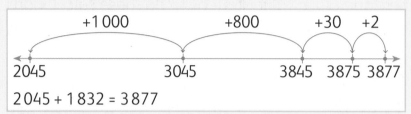

2 045 + 1 832 = 3 877

Use **expanded notation** and **place value**:

2 345 + 1 342

Round and estimate first: 2 300 + 1 300 = 3 600

2 345 + 1 342 = 2 000 + 1 000 + 300 + 300 + 40 + 40 + 5 + 2

= 3 000 + 600 + 80 + 7

= 3 687

The answer is close to the estimate, so it seems **reasonable**.

Use the column method: 1 245 + 5 827

Round and estimate first: 1 000 + 6 000 = 7 000

1 Add the ones first. **Regroup** if the **sum** is more than 9.

$$
\begin{array}{r}
1\,2\,\overset{1}{4}\,5 \\
+\,5\,8\,2\,7 \\
\hline
2
\end{array}
$$
5 + 7 = 12

2 Add the tens next.

$$
\begin{array}{r}
1\,2\,\overset{1}{4}\,5 \\
+\,5\,8\,2\,7 \\
\hline
7\,2
\end{array}
$$
1 + 4 + 2 = 7

3 Add the hundreds.

$$
\begin{array}{r}
\overset{1}{1}\,2\,\overset{1}{4}\,5 \\
+\,5\,8\,2\,7 \\
\hline
0\,7\,2
\end{array}
$$
2 + 8 = 10

4 Add the thousands.

$$
\begin{array}{r}
\overset{1}{1}\,2\,\overset{1}{4}\,5 \\
+\,5\,8\,2\,7 \\
\hline
7\,0\,7\,2
\end{array}
$$

Maths ideas

In this unit you will:
* use different methods to add numbers with up to four digits
* understand how to regroup numbers in any place
* use **rounded** numbers to find approximate answers (estimates)
* use **estimates** to decide whether an answer is reasonable
* draw models to solve problems involving addition.

Key words

addition	place value
add	reasonable
rounded	regroup
estimates	sum
expanded notation	total

Check:

The answer is close to the estimate so it seems reasonable.

1   Estimate and then add. Show your working.

   a   2 345 + 1 876   b   5 178 + 7 234   c   2 098 + 1 345
   d   3 999 + 3 081   e   8 234 + 1 876   f   4 876 + 4 654
   g   2 323 + 943   h   1 234 + 4 543   i   1 345 + 999

Explain

Bar models can help you to show the information you are given in a problem.

Example 1

A farmer planted 1 325 bean plants in one field and 3 450 cabbages in another field. How many plants is this in all?

Total?

1 325 beans	3 450 cabbages

This is 4 775 plants.

```
  1 325
+ 3 450
─────────
  4 775
```

Example 2

A farmer planted 1 250 banana trees in the north field. He planted 45 more than that in the south field. How many trees did he plant in both fields?

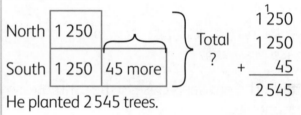

North | 1 250
South | 1 250 | 45 more

Total ?

He planted 2 545 trees.

```
  1 250
  1 250
+    45
─────────
  2 545
```

Problem-solving

2   An airline gathered this information.

Day	Mon	Tues	Wed	Thurs	Fri
Distance flown (in km)	550	323	546	457	432
Number of passengers	1 098	2 464	3 209	3 876	945

   a   What is the total distance flown in 5 days?
   b   How many passengers did the airline carry altogether on Wednesday and Thursday?
   c   On Saturday the airline carried 3 654 passengers and on Sunday it carried 92 more than on Saturday. How many passengers did it carry on these two days?

What did you learn?

1   Find the sum of 5 754 and 2 218.

2   Make up two addition problems of your own using the table above. Exchange problems with a partner. Check each other's answers.

D Subtraction

Subtraction involves finding the difference between two numbers.

You can use different methods to find the **difference** between two numbers.

Read the examples carefully. Make sure you know how to use each method.

Jumping back in chunks on a number line: 2 345 – 1 648

Estimate: 2 300 – 1 600 = 700

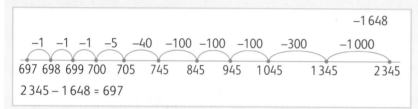

2 345 – 1 648 = 697

Check: The answer is close to the estimate, so it seems reasonable.

Expanded notation and **place value:** 3 765 – 1 283

Estimate: 3 700 – 1 300 = 2 400

3 000 – 1 000 = 2 000   **Subtract** the thousands
700 – 200 = $\overset{4}{5}$00   Subtract the hundreds
60 – 80   Borrow 100 from 500 to make more tens
160 – 80 = 80   Subtract the tens
5 – 3 = 2   Subtract the ones
2 000 + 400 + 80 + 2 = 2 482

Check: The answer is close to the estimate, so it seems reasonable.

Column methods:

3 186 – 877   Estimate first: 3 000 – 900 = 2 100

1   3 1 8 6   Write the greater number first.
    – 8 7 7   Line up the place values.
         9   Subtract the ones first.
           Regroup if necessary.
           16 – 7 = 9

2   3 1 8 6   Subtract the tens.
    – 8 7 7   7 – 7 = 0
       0 9

3   3 1 8 6   Subtract the hundreds.
    – 8 7 7   Regroup if necessary.
     3 0 9   11 – 8 = 3

4   3 1 8 6
    – 8 7 7   2 – 0 = 2 ← Subtract
   2 3 0 9   the thousands.

Explain

Look at this example. The greater number has 0 in the ones, tens and hundreds places so you have to regroup the thousands.

Work through the example carefully to make sure you understand how to do this.

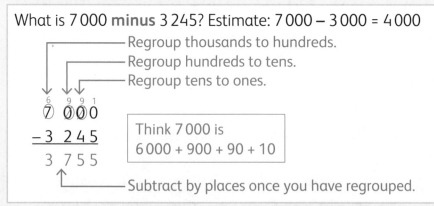

What is 7 000 **minus** 3 245? Estimate: 7 000 – 3 000 = 4 000

Regroup thousands to hundreds.
Regroup hundreds to tens.
Regroup tens to ones.

$$\begin{array}{r} {}^{6}\cancel{7}\ {}^{9}\cancel{0}{}^{9}\cancel{0}{}^{1}0 \\ -\ 3\ 2\ 4\ 5 \\ \hline 3\ 7\ 5\ 5 \end{array}$$

Think 7 000 is
6 000 + 900 + 90 + 10

Subtract by places once you have regrouped.

1 Subtract. Estimate first and show your working.

a 876 – 483	b 987 – 599	c 398 – 206
d 1 245 – 899	e 9 345 – 8 876	f 9 876 – 4 952
g 9 888 – 3 498	h 9 713 – 8 764	i 8 000 – 3 045

2 Subtract. Estimate first and show your working.

a 8 210 – 1 459	b 9 400 – 2 485	c 8 800 – 4 564
d 6 200 – 1 765	e 8 000 – 3 288	f 9 000 – 8 754
g 6 000 – 4 450	h 9 000 – 4 599	i 7 323 – 4 523

Explain

You can use bar models (like the ones you used for addition) when you need to solve subtraction problems.

Example 1

The combined area of two islands is 7 645 square kilometres. If one island has an area of 3 612 square kilometres, what is the area of the other island?

We know that 3 612 + ? = 7 645

So, 7 645 – 3 612 = ?

Total area = 7 645

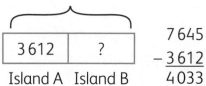

$$\begin{array}{r} 7\ 645 \\ -\ 3\ 612 \\ \hline 4\ 033 \end{array}$$

Island A   Island B

Example 2

Field A contains 2 074 banana trees. Field B has 90 fewer trees than Field A. How many trees are in Field B?

Field A – 90 = Field B

2 074 – 90 = ?

Field A = 2 074

Field B ?

90

$$\begin{array}{r} 2\ 074 \\ -\ \ \ \ 90 \\ \hline 1\ 984 \end{array}$$

27

Problem-solving

3   A large cruise ship can carry 8 500 passengers. If 7 387 passengers have already booked, how many places are left on the ship?

4   9 200 people booked cruises in the first two months of this year. This is 1 829 more than in the first two months of last year. How many people booked cruises in the first two months of last year?

5   Alicia subtracted a number from 5 000 and got an answer of 1 852. What number did she subtract?

6   Jayson wants to buy a car that costs $4 527. He has saved $3 649. How much more does he need?

7   Maria flew 4 059 km from the Caribbean to Los Angeles in the USA. Josh flew 2 329 km from the Caribbean to Mexico City in Mexico. How much further did Maria fly?

8   The table shows the points that Stacy and Cali scored in a spelling competition.

   a   Who won the competition?

   b   How many more points did the winner score than the other person?

Stages	Stacy	Cali
1st round	95	115
2nd round	105	75
3rd round	60	65

9   Nitha has 2 106 beads. Ella has 1 003 fewer than Nitha and Stacy has 309 fewer than Nitha and Ella combined.

   a   How many beads does Stacy have?

   b   How many beads do the three girls have altogether?

   c   How many more do they need to have 9 999 beads between them?

10   Two countries have a combined area of 8 452 square kilometres. One country is 1 246 square kilometres larger than the other. What is the area of each country?

What did you learn?

1   Subtract. Estimate first and show all your working.

   a   3 245 – 1 245   b   4 009 – 2 389   c   5 000 – 2 040

2   Denton has 2 452 points in a game. His sister has 3 124 points. They combine their points and give up 4 500 points to win a prize. How many points will they have left?

Topic 3 Review

Key ideas and concepts

Copy and complete these sentences to summarise what you learnt in this topic.

1   Basic number facts are important because _____.
2   Addition means you have to _____.
3   Subtraction means you have to _____.
4   Some of the methods you can use to add or subtract are _____.
5   A number line is useful for _____.
6   Regrouping numbers means _____.
7   When you have to solve problems it is important to _____.

Think, talk, write ...

1   Work in pairs. Pretend you are teachers.
    *   How would you teach the difference between adding and subtracting?
    *   Show the class how to work out 2 000 − 345.

Quick check

1   Write down the year we are in. Add this to the year we were in last year.

2   Add. Use mental strategies where possible.
    a   29 + 106   b   432 + 575   c   1782 + 324
    d   3039 + 152   e   2005 + 409   f   6092 + 1949

3   Write each of these numbers in numerals.
    a   four hundred and thirty-seven
    b   one thousand eight hundred and twenty-nine
    c   twelve thousand eight hundred
    d   twenty-three thousand four hundred and two

4   Calculate.
    a   1991 + 234 + 4568   b   1285 − 873   c   8403 − 3460
    d   3819 + 4214   e   9876 − 4388   f   4000 − 399

5   The sum of three numbers is 2 345. Write four different addition sums that will give this result.

6   The difference between two four-digit numbers is 879. What could the numbers be?

29

A

This building is made of metal and glass. What different shapes have the architects used in the design? Find examples of shapes with four, five and six sides. Show them to your partner.

Teaching notes

Classifying shapes

* Shapes can be classified and grouped using their properties.
* Properties include the number and length of sides and the number and size of angles.

Polygons

* *Poly-* means many. Polygons are closed 2-D shapes with many sides.
* Polygons are named according to the number of sides they have.
* To make a closed shape, you need at least three sides, so the triangle is the first polygon.
* All polygons with four sides are called quadrilaterals.
* Groups of polygons such as triangles and quadrilaterals can be divided into smaller groups using the properties of the shapes. For example, equilateral triangles are those with all sides and angles equal. Squares are quadrilaterals with four equal sides and four right angles.

Look at the pattern on this phone screen. What shapes does it use? How are the shapes all the same? How are they different? How many shapes can you count? Compare your answer with your partner's answer.

Think, talk and write

A **Classifying shapes**
(pages 32–33)

1 Think of all the shapes you already know. Look around the classroom and take turns to show and name different shapes.

2 Find the two closed shapes in the classroom with the least and most sides. Draw them in your book.

B **More about triangles and quadrilaterals** *(pages 34–36)*

1 The prefix *tri-* means three. What do you think these words mean?
   a tricycle
   b triathlon
   c triplets

2 What does the prefix *quad-* mean? Can you find some words that start with quad-? What do they mean?

A Classifying shapes

An **open** shape has ends that do not meet.

A **closed** shape does not have any loose ends.
All the sides meet up.

Closed shapes that have only straight **sides** are
called **polygons**.

A polygon has an equal number of sides and angles.

A **triangle** is the polygon with the least number of sides
and angles.

The table shows you the names as well as examples of
different polygons.

Maths ideas

In this unit you will:
* classify shapes using
  their properties
* learn the names of
  some groups of shapes
* **compare** squares and
  rectangles to see how
  they are the same and
  how they differ.

Key words

open	polygons
closed	triangle
compare	squares
sides	rectangles

Type of polygon	Number of sides	Number of angles	Example
Triangle	3	3	
Quadrilateral	4	4	
Pentagon	5	5	
Hexagon	6	6	
Heptagon	7	7	
Octagon	8	8	
Nonagon	9	9	
Decagon	10	10	

1 Divide these shapes into two groups: polygons and not polygons.

2 Look at these groups of shapes.

Group A   Group B   Group C   Group D   Group E

  a   What properties do you think the student used to place the shapes in each group?

  b   How could you divide the shapes in group E further? Explain your choices.

3 Name each of these polygons. Refer to the table if you need to.

a   b   c   d   e

4 Read the information in the box carefully.

  ✳ **Squares** have four sides and four angles. The four sides are equal in length. The four angles are all right angles.

  ✳ **Rectangles** have four sides and four angles. The two pairs of sides opposite each other are equal. The four angles are all right angles.

  a   Which of these shapes is the square?

  b   Jolene says both shapes can be called rectangles. Is she correct? Why?

 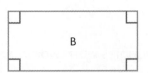

  c   Teresa says they can both be called squares. Is she correct? Why?

What did you learn?

1 Look at the football carefully.

  a   What is the mathematical name of the white shapes?

  b   What is the mathematical name of the black shapes?

2 Explain how you can tell whether a quadrilateral is a square or a rectangle.

B More about triangles and quadrilaterals

All shapes with three angles and three sides are **triangles**, but there are different types of triangles.

Triangles can be sorted into groups using the size of their angles, the lengths of their sides or a combination of these two things.

Right-angled triangles

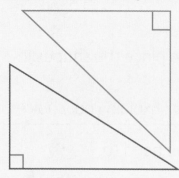

Triangles in which all angles are smaller than a right angle

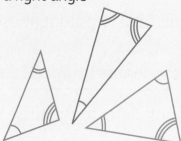

Triangles that contain an angle that is greater than a right angle

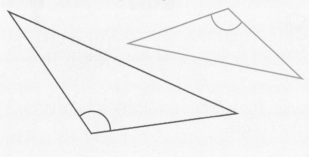

Triangles whose three sides are equal in length

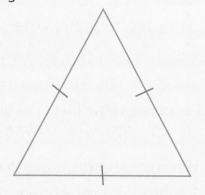

Triangles with two sides equal in length

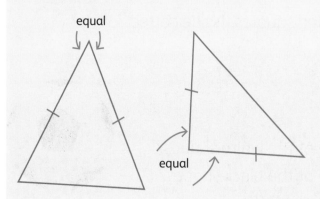

equal

Triangles with no sides equal in length

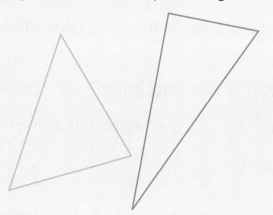

equal

Explain

All shapes with four sides are **quadrilaterals**, but there are different types of quadrilateral. You already know about squares and rectangles and their special properties. Here are four more types of quadrilateral, each with its own special properties:

This is a **parallelogram**. It has two pairs of opposite sides that are parallel and equal in length.

A parallelogram is like a pushed-over rectangle.

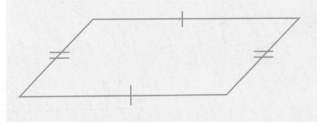

This is a **rhombus**. It has four equal sides.

A rhombus is like a pushed-over square.

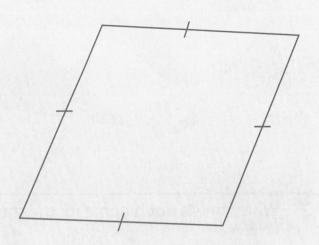

This is a **kite**. It has two pairs of equal sides. The equal sides are next to each other.

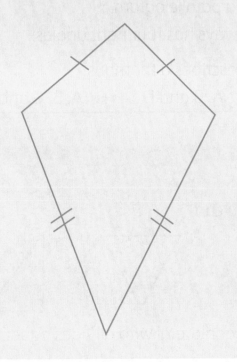

This is a **trapezium**. It has one pair of opposite sides that are the same distance apart all along their length.

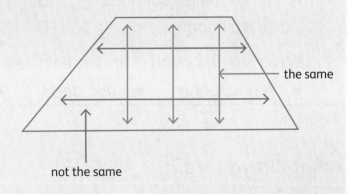

the same

not the same

1   **Measure** the sides of these triangles and look at the angles to sort them into three or more groups. Write a statement to describe the triangles in each group.

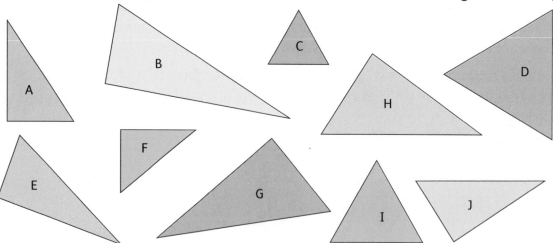

Problem-solving

2   Which one is **not** a name for the shape below?

   * polygon   * quadrilateral
   * parallelogram   * trapezium

3   Read these statements.

   A It has four equal sides.   **B** It is a parallelogram.
   C It is a quadrilateral.   **D** It always has four right angles.

   Which combination of statements best describes a rhombus?
   * A and C only   * A, B and C   * A, B and D   * A, B, C and D

What did you learn?

1   Describe each triangle as accurately as you can.

   a   b   c   d   e

2   Winston's teacher asked him to draw a quadrilateral with all sides equal. Which of these shapes could he draw?

   * square   * rectangle   * parallelogram
   * kite   * rhombus   * trapezium

Topic 4 Review

Key ideas and concepts

Write notes about each picture to summarise what you learnt in this topic.

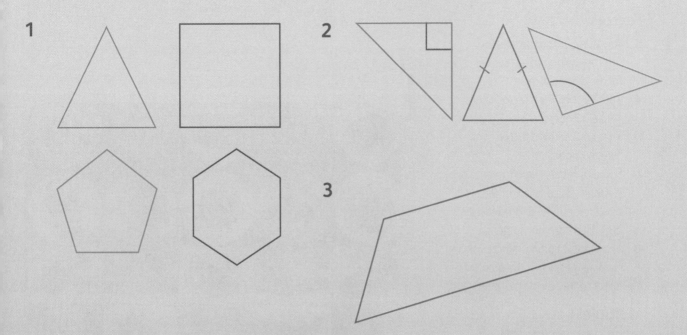

Think, talk, write ...

Work in pairs. Imagine you've been asked to write a page for a children's mathematics dictionary. The topic of your page is polygons. Think carefully about what information you need to include. Then plan and draw up an illustrated page on this topic.

Quick check

1   Give the correct name for each of these polygons.

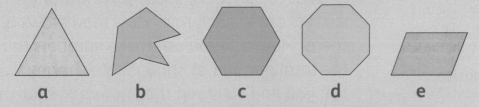

a	b	c	d	e

2   Draw a right-angled triangle with two equal sides.

3   Why can you say that a rectangle is a special type of parallelogram?

4   Why can you say that a square is a special type of rectangle?

5   Can you also say that a square is a type of:

   a   parallelogram?   b   rhombus?

Teaching notes

Factors

* Any number that divides into another without a remainder is a factor of that number.
* The number 1 is a factor of all whole numbers.
* The number 2 is a factor of all even numbers.
* Some numbers share factors. For example, 3 is a factor of 9 as well as a factor of 12. Shared factors are called common factors.
* The highest common factor of two or more numbers is the greatest factor shared by the numbers.

Multiples

* When you multiply a whole number by any other number, the product is a multiple of that number. For example, $5 \times 1 = 5, 5 \times 2 = 10, 5 \times 3 = 15$. The products 5, 10 and 15 are all multiples of 5.
* 10 is a multiple of 5, but it is also a multiple of 10 ($10 \times 1 = 10$). We say 10 is a common multiple.
* 10 is the lowest number that is both a multiple of 5 and a multiple of 10, so it is the lowest common multiple of the two sets.

Prime and composite numbers

* Prime numbers have only two factors: 1 and the number itself.
* Composite numbers have more than two factors.
* The number 1 is neither prime nor composite.
* Most prime numbers are odd. The only even prime number is 2.

A

Look at the sheet of stickers. How many stickers are in each row? How many rows are there? How can you use these numbers to find the total number of stickers? How many ways can you find to group the stickers so that there is the same number in each group and none left over?

B

At a fete, every fourth person who enters is given a free apple and every sixth person is given a free juice. Try to work out which person would be the first one to receive both. When the organisers have given out 32 apples, how many people have entered?

C

There is only one way to arrange three counters in a rectangular array. Is there more than one way to arrange four counters? Arrange 5 and 6 into rectangular arrays. How many ways can you find for each number?

Think, talk and write

A Factors and multiples
(pages 40-41)

1 Make a list of all the whole numbers that can divide into 24 without leaving a remainder.

2 Read this information and discuss it with your partner.

> These are multiples of 5:
> 5, 10, 15, 20, 55, 90, 100
> These are not multiples of 5:
> 1, 12, 18, 24, 62, 99

Write in your own words what you think a multiple is.

B Common factors and multiples *(pages 42-43)*

1 Which numbers are factors of both 10 and 20? List them.

2 Ella exercises every second day. Maria exercises every third day. If they both exercise on Monday, on which day will they next exercise together?

C Different types of numbers
(page 44)

1 Find all the possible pairs of whole numbers you can multiply to get each of these products:

7   11   17   23

What do you notice?

2 How can you tell whether a number is odd or even?

A Factors and multiples

Explain

A **factor** of a number is any number that divides into it without a remainder.

2 is a factor of 6 because $6 \div 2 = 3$ (you know that $2 \times 3 = 6$).

4 is not a factor of 6 because $6 \div 4 = 1$ remainder 2. You cannot **multiply** 4 by a whole number to get a **product** of 6.

There are three different pairs of factors that give a product of 12.

$1 \times 12 = 12$

$2 \times 6 = 12$

$3 \times 4 = 12$

Maths ideas

In this unit you will:
* learn about factors and multiples of numbers
* list the factors of different numbers
* find multiples of a given number.

Key words

factor	product
multiply	multiples

This shows that a number can have many factors.

1, 2, 3, 4, 6 and 12 are all factors of 12.

This means that 12 can be divided by each of these numbers without a remainder.

All numbers can be divided by 1, so 1 is a factor of every number.

All numbers can also be divided by themselves (and the result will be 1), so every number is also a factor of itself.

To find all the pairs of factors of a number you can build a multiplication tree.

Example 1	**Example 2**
Factors of 8	Factors of 30
$8 \begin{cases} 1 \times 8 \\ 2 \times 4 \end{cases}$	$30 \begin{cases} 1 \times 30 \\ 2 \times 15 \\ 3 \times 10 \\ 5 \times 6 \end{cases}$
Factors are 1, 2, 4, 8	Factors are 1, 2, 3, 5, 6, 10, 15, 30

List the factors in ascending order and don't repeat any.

Multiples of a number are found when you multiply that number by any other whole number.

The first five multiples of 3 are:

3	6	9	12	15
(3×1)	(3×2)	(3×3)	(3×4)	(3×5)

The first ten multiples of 4 are: 4, 8, 12, 16, 20, 24, 28, 32, 36, 40

1   Read each statement. Say whether it is true or false.
  a   5 is a factor of 20.
  b   5 is a factor of 24.
  c   9 is a multiple of 3.
  d   All even numbers are multiples of 2.
  e   10 is a factor of 80.
  f   75 is a multiple of 10.

2   List all the factors of each number.
  a   26   b   46   c   32   d   25   e   50   f   40
  g   63   h   55   i   49   j   27   k   35   l   14

3   Which of these numbers are multiples of 3?
    12   16   21   24   25

4   Which of these numbers are multiples of 7?
    14   21   24   28   30   37   42

5   Write the first five multiples of each number.
  a   5   b   9   c   7   d   10   e   2

6   List the multiples of:
  a   4 between 25 and 45   b   100 between 450 and 950.

Problem-solving

7   Shawnae has five number cards. Her friends each
    choose a card.
    * Jayson says his card is a factor of 3 and a factor of 12.
    * Nisha says her card is not a factor of 12 or 14.
    * Marie says her card is a factor of 18.
    * Marten says his card is a factor of 8.
    Which card is Shawnae left with?

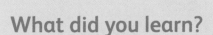

3   4   5   6   7

What did you learn?

1   What are the factors of 21?

2   How do you know that 99 is a multiple of 9, but not a multiple of 10?

3   Write five numbers between 1 396 and 1 424 that are multiples of five.

B Common factors and multiples

Explain

Some numbers share factors.

Factors of 12: 1, 2, 3, 4, 6, 12
Factors of 16: 1, 2, 4, 8, 16

1, 2 and 4 are factors of 12 and factors of 16.

We call these **common** factors.

4 is the greatest number that is common to both sets. We call this the **highest common factor (HCF)** of 12 and 16.

HCF is a short way of writing highest common factor.

To find the HCF, list the factors of both numbers. Circle the common factors and find the highest number that is common to the sets.

Example 1

Find the highest common factor of 24 and 36.
Factors of 24: ①, ②, ③, ④, ⑥, 8, ⑫, 24
Factors of 36: ①, ②, ③, ④, ⑥, 9, ⑫, 18, 36
The HCF of 24 and 36 is 12.

The first five multiples of 3 and 4 are listed here:
Multiples of 3: 3, 6, 9, ⑫, 15
Multiples of 4: 4, 8, ⑫, 16, 20

The lowest number that is common to both sets is 12.

We say that 12 is the **lowest common multiple (LCM)** of 3 and 4.

To find the lowest common multiple of two numbers, list the multiples in order till you get to one that is shared by both.

Example 2

Find the LCM of 12 and 3.

Multiples of 12: ⑫, 24, 36

Multiples of 3: 3, 6, 9, ⑫
You can stop when you reach a common multiple.

The LCM of 12 and 3 is 12.

Maths ideas

In this unit you will:
* list sets of factors and multiples to find the lowest common multiple and highest common factor
* use lowest common multiples and highest common factors to solve problems.

Key words

common

highest common factor (HCF)

lowest common multiple (LCM)

1   Find the lowest common multiple of each set of numbers.

a   6 and 4   b   2 and 9   c   12 and 8

d   2 and 7   e   5 and 8   f   6 and 7

g   2, 3 and 4   h   4, 6 and 10   i   3, 7 and 8

2   Find the highest common factor of each set of numbers.
  a   15 and 20   b   12 and 16   c   6 and 9
  d   28 and 42   e   32 and 36   f   24 and 48
  g   12, 15 and 18   h   14, 21 and 30   i   40, 80 and 120

3   Look at this subtraction model for finding the highest common factor of two
    numbers without listing all the factors.

Work on the side of the bigger number.

60	84
60	84 – 60
60 – 24 = 36	24
36 – 24 = 12	24
12	24 – 12 = 12

Subtract the smaller number from the bigger number.
60 > 24, so subtract 60 – 24
36 > 24, so subtract 36 – 24
24 > 12, so subtract 24 – 12
Stop when the numbers are equal.

Try this method to find the highest common factor of:
  a   40 and 70   b   100 and 64   c   72 and 20

Problem-solving

4   Paper plates are sold in packs of 8 and paper cups are
    sold in packs of 6. What is the fewest packs of plates
    and cups you could buy to get the same number of cups and plates?

5   Miley has 30 red beads and 48 blue beads to put onto bangles. She wants
    to divide the beads so that each bangle has the same number of red and
    blue beads.
  a   What is the fewest number of bangles she can make?
  b   How many red beads will there be on each bangle?
  c   How many blue beads will there be on each bangle?

6   The lowest common multiple of three numbers is 18. What could the
    numbers be?

What did you learn?

1   What is the HCF of 12 and 42?   2   What is the LCM of 5 and 6?

3   Is it possible to find the highest common multiple of two numbers? Explain
    your answer.

C Different types of numbers

Explain

Numbers that have more than two factors are called **composite** numbers.
* 25 is a composite number. Its factors are: 1, 5 and 25.
* 12 is a composite number. Its factors are: 1, 2, 3, 4, 6, 12.

Numbers that have only two factors are called **prime** numbers.
* A prime number can only be divided by 1 and itself.
* 7 is a prime number. Its factors are 1 and 7.
* 11 is a prime number. Its factors are 1 and 11.

The number 1 is a special case. 1 is not a prime number nor a composite number, because it only has one factor.

The number 1 is an **odd** number. All numbers with 1, 3, 5, 7 or 9 in the units place are odd numbers.

Numbers with 0, 2, 4, 6 or 8 in the units place are **even** numbers.

Composite numbers can be odd or even. The only even prime number is 2, all the others are odd.

Maths ideas

In this unit you will:
* use what you already know about odd and even numbers
* understand the difference between prime numbers and composite numbers
* group and classify different types of numbers.

Key words

composite	even
prime	classify
odd	

1   The first five prime numbers are listed here. List the next five.

2, 3, 5, 7, 11

2   The first five composite numbers are listed here. List the next five.

4, 6, 8, 9, 10

3   **Classify** each of these numbers as odd or even and prime or composite.

11   15   18   21   30   37   39   41   43   45

Problem-solving

4   Which prime numbers can be added to get a total of 20? Can you find more than one combination?

What did you learn?

1   List the odd prime numbers less than 20.

2   List the composite numbers between 10 and 25.

3   Which two prime numbers are factors of 20?

Topic 5 Review

Key ideas and concepts

Match the mathematical terms on the left to the meanings on the right to summarise what you learnt in this topic.

Factor	Numbers that have more than two factors
Even	Numbers that can be divided by 2 with no remainder
Multiples	A number that will divide into another with no remainder
Prime	The highest number common to sets of factors
Odd	Numbers with 1, 3, 5, 7 or 9 in the units' position
HCF	The lowest number in two or more sets of multiples
LCM	A number that can only be divided equally by itself and 1
Composite	Products of a number and other numbers

Think, talk, write …

1   Why are there no even prime numbers greater than 2?

2   Is it possible for two numbers to have no common factors at all? Explain your answer.

Quick check

1   a   List the factors of 40.

   b   List the factors of 48.

   c   List all the common factors of 40 and 48.

   d   Which of the common factors are prime numbers?

2   a   List the first ten multiples of 6.

   b   List the first ten multiples of 8.

   c   What is the lowest common multiple of 6 and 8?

3   Here is a set of numbers: | 2   5   7   11   12   15   20   36   41 |

   a   List the odd numbers.

   b   List the even numbers.

   c   List the prime numbers.

   d   List the composite numbers.

4   I am thinking of a number. It is an odd factor of 39. It is greater than 11, but smaller than 25. What is the number?

Topic 6 Computation (2)

Teaching notes

Multiplication and division facts

∗ Students should know their times tables very well. Continue to skip count in multiples and to drill and practice these regularly. They will use these facts over and over when they multiply and divide in a higher number range.

∗ The inverse relationship between multiplication and division allows students to use multiplication facts to find related division facts. Continue to generate and use fact families to reinforce this.

Multiplication

∗ Students will now move to pen-and-paper methods of multiplying higher numbers. It is important to continue to model multiple strategies for doing this and to allow students to move towards a more formal algorithm only when they are ready to do so.

Division

∗ At this level, students only divide by single-digit numbers, so they can use their known facts to solve most division problems.

∗ Short division is an important step in developing good understanding. Students need lots of practice in this method to build foundations for later work with higher numbers.

Order of operations

∗ There is an agreed set of rules for how to work when there is more than one operation in a number sentence. Students need to know the rules and be able to apply them.

∗ The rules state that any operations in brackets are done first, no matter where they appear in the number sentence.

∗ Once the brackets are cleared, multiplications and divisions are done in order from left to right.

∗ Addition and subtraction are done last, also from left to right.

∗ Most modern calculators, including those in mobile phones, apply these rules automatically. Students should experiment to see whether their own calculator does this or not.

A

The mathematical word for an arrangement of rows and columns like this is an array. Use the array in the photograph to make up one division, one multiplication, one addition and one subtraction number sentence. Compare your sentences with a partner's.

B

Count the eggs in the box. What multiplication facts give you this product? What fact family can you make for this set of eggs? How would you work out the number of eggs in 2 boxes? In 9 boxes?

1234

A load of 300 peppers was delivered to the market. The market owner wants to sell them in bags of five. He knows that 5 × 6 is 30. How can this help him work out how many bags he needs for the 300 peppers?

Calculators are programmed to do calculations, but they don't all work in the same way. Enter 5 + 3 × 4 into any calculator. What answer do you get? If your calculator says 17, it has applied the rules of mathematics. If your calculator says 34, it has worked from left to right and not applied any rules.

Think, talk and write

A Number facts (pages 48–49)

Look at the diagram.

1 What is the name of this type of diagram?

2 How can this diagram help you multiply and divide?

3 Use the diagram to write the fact family for 6, 7 and 42.

B Multiplication (pages 50–51)

1 A farmer harvests enough sugar cane to make 254 sacks of sugar each week.

   a Can you work out how much he harvests in 4 weeks by adding? How?

   b How would you multiply to find the answer?

2 How can writing 46 as 40 + 6 make it easier for you to multiply it by 5?

C Division (pages 52–53)

1 Nicky knows that 3 × 6 = 18. How can this help her share 180 pencils equally among three classes?

2 32 ÷ 4 = 8

   a How can you use this fact to find 320 ÷ 4?

   b What is 320 ÷ 8?

D Order of operations (pages 54–56)

Two girls worked on this problem: Josh has 3 five-dollar bills and two dollar coins. How much money does he have?

1 Which is the correct answer?

2 What did the other girl do incorrectly?

A Number facts

Explain

Last year you learnt your times tables from memory and you used these tables to work out other **multiplication** and **division** facts.

Read the examples to make sure you remember how multiplication and division are related to each other.

$4 \times 3 = 12$

The answer to a multiplication is called a **product**.
Multiply the **factors** to get a product.

$$
\begin{array}{ccc}
4 & \times \; 3 & = \; 12 \\
\searrow & \swarrow & \downarrow \\
\text{factors} & & \text{product}
\end{array}
$$

5, 10, 15, 20, 25   These are multiples of 5.
Each number in a skip counting pattern is a **multiple**.

$4 \times 6 = 24$ Operations that undo each other are called **inverse** operations.

$24 \div 6 = 4$ Division is the inverse of multiplication.

There are two ways to write division sentences.

$$
\begin{array}{ccccc}
24 & \div & 6 & = & 4 \\
\downarrow & & \downarrow & & \downarrow \\
\textbf{dividend} & & \textbf{divisor} & & \textbf{quotient}
\end{array}
$$

divisor ↘   ↗ quotient
$$\frac{4}{6\overline{)24}}$$
dividend ←

A **fact family** shows all the related multiplication and division facts for a set of three numbers.

If you know one fact, you can use it to work out all the others.

Example

$4 \times 7 = 28$	$7 \times 4 = 28$	$28 \div 7 = 4$	$28 \div 4 = 7$

Maths ideas

In this unit you will:
* revise the multiplication and division facts you already know
* use known facts to work out other facts
* improve your memorisation of times tables.

Key words

multiplication
division
product
factors
multiple
inverse
dividend
divisor
quotient
fact family

Think and talk

What do you do to help you remember your times tables? Share your ideas with your group.

1   Multiply. Write the product only.

a	2×8	b	3×7	c	8×7	d	5×7	e	4×6
f	5×5	g	4×4	h	2×7	i	9×1	j	8×9
k	3×9	l	6×9	m	9×3	n	5×8	o	8×8

2 Write the multiplication and division fact family for each set of numbers.

a 3, 6, 18   b 2, 9, 18   c 5, 9, 45   d 7, 6, 42

e 3, 9, 27   f 1, 9, 9   g 7, 9, 63   h 5, 6, 30

i 7, 8, 56   j 2, 4, 8   k 6, 6, 36   l 8, 9, 72

3 Write a fact family that has 7 as one of the factors and a product of 28.

4 Divide. Write the quotient only.

a $27 \div 3$ b $24 \div 8$ c $36 \div 6$ d $5\overline{)35}$

e $16 \div 2$ f $24 \div 3$ g $16 \div 4$ h $8\overline{)72}$

i $30 \div 6$ j $49 \div 7$ k $8\overline{)32}$ l $2\overline{)18}$

Problem-solving

5 Conchs are tied up in bunches of 5.

a How many conchs are there in 8 bunches?

b How many bunches of 5 can you make if there are 45 conchs altogether?

6 A farmer has 48 mangoes.

a How many packs of 4 can she make?

b If there are 8 packs with an equal number of mangoes, how many are there in each pack?

c If she sells 6 packs for $7.00 each, how much money will she receive?

7 A cap costs $6.00 at a market stall.

a Tony buys a cap for each of his 8 friends. How much will they cost altogether?

b Andy spent $30.00 on caps. How many caps did he buy?

c Nisha buys 4 caps and pays with 3 ten-dollar bills. How much change will she get?

What did you learn?

1 Write the fact family for each set of numbers.

a 3, 9, 27   b 7, 4, 28   c 4, 10, 40

2 Sandy practices the violin for 2 hours a day, 6 days a week.

a How long does she practice in 1 week?

b How many hours will she practice in 3 weeks?

B Multiplication

Explain

Do you remember the different pen-and-paper methods you can use to multiply larger numbers?

You can use the methods you have already learnt, to multiply any numbers.

These examples show you how to use the different methods to multiply by a two-digit number.

Multiply 58 by 23:

Estimate: $60 \times 20 = 60 \times 2 \times 10 = 120 \times 10 = 1\,200$

Use expanded notation:

$58 \times 23 = (58 \times 20) + (58 \times 3)$ **Expand** one **factor** first.

$58 \times 20 = 58 \times 2 \times 10 = 116 \times 10 = \quad 1\,160$

$58 \times 3 = 50 \times 3 + 8 \times 3 = 150 + 24 = \quad \underline{\quad 174}$

$\qquad\qquad\qquad\qquad\qquad\qquad\qquad 1\,334$

Add the partial products.

Use a **grid:**

Expand both numbers.

×	20	3
50	1 000	150	→ 1 ^{1}150
8	160	24	→ 184

Add across.   1 334

This is the **product.**

Use a **column** method:

$$
\begin{array}{r}
58 \\
\times\ 23 \\
\hline
24 \quad 3 \times 8 \\
{}^1150 \quad 3 \times 50 \\
160 \quad 20 \times 8 \\
\underline{1\,000} \quad 20 \times 50 \\
1\,334 \text{ Add the partial products.}
\end{array}
$$

Maths ideas

In this unit you will:
* use pen-and-paper methods to multiply larger numbers
* solve problems that involve multiplication.

Key words

estimate

expand

factor

grid

column

product

---

1   Use the method you find the easiest to find these products. Estimate first and show your working.

a 23×8 b 49×8

c 98×9 d 92×6

e 123×5 f 435×7

g 129×9 h 452×8

i 234×3 j 987×2

k 4×342 l 9×265

2   Look at these two multiplications. The factors are missing in each, but the product is given. The missing digits are 2, 3 and 4 in both calculations. Work out what the factors are in each.

$\square\square \times \square = 86$ $\square \times \square\square = 126$

3   Multiply. Estimate first. Then use the method you find the easiest.

a 27×13	b 37×22	c 25×15	d 27×32
e 17×54	f 64×32	g 29×82	h 32×19
i 87×16	j 23×12	k 43×30	l 90×29

4   This is Sherry's homework. Without doing the calculations, decide whether each answer is reasonable or not.

a $21 \times 38 = 798$	b $34 \times 18 = 212$	c $17 \times 23 = 391$
d $33 \times 26 = 828$	e $20 \times 26 = 480$	f $18 \times 32 = 600$

5   Find the product of: a   12 and 85   b   23 and 49.

6   Nick bought 23 pencils for 45 cents each. What did he pay?

7   Sharon runs 15 kilometres every week. How far will she run in a year?

8   Each classroom in a school has 28 chairs. If the school has 24 classrooms, how many chairs are there?

9   How many banana plants are there in 25 rows if there are 45 plants in each row?

10   Janae did this calculation, but she rubbed out two digits by mistake. What are they?

$$\begin{array}{r} 146 \\ \times\ \ 3 \\ \hline 7\ 0 \end{array}$$

What did you learn?

Calculate.

1 123×3	2 76×5	3 129×9
4 32×28	5 19×45	6 12×96

C Division

When the number you are dividing into (the **dividend**) is not in the times table, you can use written methods to find the **quotient**. One method is called short **division**.

Example 1

$$\begin{array}{r} 1 \\ 7\,\overline{\smash{)}\,8\,{}^1\!4} \end{array}$$

Start with the tens.
$8 \div 7 = 1\ r\ 1$

You have 1 left over.
Carry to the ones column.

$$\begin{array}{r} 1\ \ 2 \\ 7\,\overline{\smash{)}\,8\,{}^1\!4} \end{array}$$

Divide the ones.
$14 \div 7 = 2$

When the **divisor** does not go exactly into the dividend you are left with a **remainder**. This example shows what to do when there is a remainder.

Example 2

$$\begin{array}{r} 1 \\ 6\,\overline{\smash{)}\,7\,{}^1\!7} \end{array}$$

$7 \div 6 = 1\ r\ 1$
Carry to the tens column.

$$\begin{array}{r} 1\ \ 2r5 \\ 6\,\overline{\smash{)}\,7\,{}^1\!7} \end{array}$$

$17 \div 6 = 2\ r\ 5$
Write the remainder as part of the quotient.

Example 3

$$\begin{array}{r} 1 \\ 4\,\overline{\smash{)}\,6\,{}^2\!24} \end{array}$$

$6 \div 4 = 1\ r\ 2$
Carry to the tens column.

$$\begin{array}{r} 1\ \ 5 \\ 4\,\overline{\smash{)}\,6\,{}^2\!2\,{}^2\!4} \end{array}$$

$22 \div 4 = 5\ r\ 2$
Carry to the ones column.

$$\begin{array}{r} 1\ \ 5\ \ 6 \\ 4\,\overline{\smash{)}\,6\,{}^2\!2\,{}^2\!4} \end{array}$$

$24 \div 4 = 6$

You can use short division methods to divide numbers in the hundreds too.

Sometimes you cannot divide the first digit, as it is smaller than the divisor. This example shows you what to do if the digit in the hundreds place is not large enough to divide.

Example 4

$178 \div 3$

$$\begin{array}{r} 5 \\ 3\,\overline{\smash{)}\,17\,{}^2\!8} \end{array}$$

$17 \div 3 = 5\ r\ 2$

1 is too small to be divided by 3, so work with 17. Write the answer above the 7.

$$\begin{array}{r} 5\ \ 9r1 \\ 3\,\overline{\smash{)}\,17\,{}^2\!8} \end{array}$$

$28 \div 3 = 9\ r\ 1$

Write the remainder as part of the quotient.

Maths ideas

In this unit you will:
* use pen-and-paper methods to perform division
* solve problems that involve division.

Key words

dividend	divide
quotient	divisor
division	remainder

1   Divide. Show your working.

 a $4\overline{)63}$ b $3\overline{)51}$ c $5\overline{)75}$ d $6\overline{)84}$

 e $3\overline{)67}$ f $4\overline{)93}$ g $7\overline{)88}$ h $8\overline{)96}$

2   Divide. Show your working.

 a $232 \div 2$ b $487 \div 5$ c $423 \div 3$ d $904 \div 8$

 e $640 \div 5$ f $228 \div 2$ g $812 \div 3$ h $413 \div 3$

3   Divide. Show your working.

 a $4\overline{)500}$ b $9\overline{)288}$ c $4\overline{)156}$ d $5\overline{)650}$

 e $8\overline{)104}$ f $3\overline{)172}$ g $6\overline{)636}$ h $3\overline{)423}$

Problem-solving

4   Pedro collects 94 eggs from his granny's chicken run.
 He puts the eggs into boxes. Each box holds 6 eggs.
 a How many boxes can he fill?
 b How many eggs will he have left over?

5   Sharon recorded how long she exercised in 1 week and found she had
 exercised for 508 minutes. How many minutes is this per day?

6   A florist has 376 roses. He sells them in bunches of nine. How many
 bunches can he make?

7   159 children are going on a school trip to a museum.
 a The teachers put the children into three groups. How many children
  are in each group?
 b The buses that will transport the children can each carry 28 children.
  Will five buses be enough?
 c Each child paid $4.00. How much money is this altogether?
 d At the museum, the children enter in groups of eight. How many
  groups will there be?

What did you learn?

1   What is the quotient when 75 is divided by 3?

2   Share 123 pencils equally among eight children. How many are left over?

3   Tom washes his car every 5 days. How many times will he wash it in 8 weeks?

D Order of operations

Mathematicians have agreed that when there is more than one operation in the same number sentence, you have to do them in a particular order to avoid confusion when you get different answers.

The order of **operations** rules state that you always do the operations inside **brackets** first. The brackets are placed around part of the calculation to **group** operations.

$2 \times (3 + 4)$ This means you **add** $3 + 4$ and then
$= 2 \times 7$ **multiply** the result by 2.
$= 14$

There may be more than one set of brackets in a calculation.

$(4 + 4) \times (12 - 10)$ Do the calculation in each bracket
$= 8 \times 2$ before you multiply the results.
$= 16$

Maths ideas

In this unit you will:
* learn about brackets and how they are used in mathematics
* apply the correct rules to do calculations when there is more than one operation.

Key words

operations
brackets
group
add
multiply

1   Do these calculations correctly. Show the steps in your working out.

    a $(3 + 3) \times 10$ b $(18 - 3) \div 5$ c $25 - (5 + 7)$ d $8 \times (4 + 2)$

    e $(3 + 4) \times 7$ f $(20 - 12) \div 8$ g $12 + (42 \div 7)$ h $(20 - 4) \div 4$

    i $7 \times (11 - 6)$ j $(3 - 2) \times 4$ k $(6 + 7) \times 3$ l $(12 - 8) \times 9$

2   Calculate correctly. Show the steps in your working out.

    a $(3 + 3) \times (14 - 4)$ b $(9 - 5) \times (2 + 7)$ c $(4 + 16) \div (12 - 7)$

    d $(26 + 4) - (3 \times 3)$ e $(10 \times 10) \div (5 + 5)$ f $(3 - 2) \times (8 + 4)$

    g $(4 \times 6) + (18 \div 6)$ h $(9 - 5) \times (19 - 10)$ i $(17 - 8) \div (3 \times 3)$

    j $(4 + 8) \times (17 - 16)$ k $(14 - 13) + (20 \div 20)$ l $(7 + 8) \times (12 \div 12)$

Problem-solving

3   Write a number sentence with brackets to represent each problem correctly. Solve it and write the answers.

    a   I have 6 packets with 5 markers in each and 3 loose markers. How many do I have in all?

    b   Subtract the quotient of 21 divided by 3 from 19. How much is left?

    c   Tickets for a show cost \$12.00. If you buy four tickets, you get \$3.00 off each price. How much will 4 tickets cost you?

Explain

When there are no brackets, you have to apply a set of rules to decide which part of the calculation to do first.

The rules for working are very simple. You already know the first rule!
* Do operations inside brackets first.
* Next, multiply or divide in order from left to right.
* Only then add or subtract in order from left to right.

When you apply these rules, there is only one correct answer, because there is only one correct order of working.

Example 1

$4 \times (4-2) + 1$ Brackets first: $4 - 2 = 2$

$= 4 \times 2 + 1$ Multiply next: $4 \times 2 = 8$

$= 8 + 1$ Then add: $8 + 1 = 9$

$= 9$

Example 2

$2 + 18 \div 3 \times 7$ Divide before you add: $18 \div 3 = 6$

$= 2 + 6 \times 7$ Multiply before you add: $6 \times 7 = 42$

$= 2 + 42$ Add: $2 + 42 = 44$

$= 44$

4   Read the statements. Say whether they are true or false.
  a For $2 + 4 \times 5$, we would do $2 + 4$ first.
  b For $(2 + 4) \times 5$, we would do $2 + 4$ first.
  c For $10 - 4 \times 2$, we would do 4×2 first.
  d For $6 + 12 \div 3$, we divide before we add.
  e For $5 \times 10 \div 2 + 3$, we do $10 \div 2$ first.
  f For $20 - 6 \times 3 + 15$, we would subtract first.

5   Solve each calculation in Question 4 correctly.

6   Apply the rules for order of operations to find the answers to each calculation.
  a $(8 - 2) + 4$
  b $18 - 4 \times 2 - 3$
  c $(12 - 9) \times (24 - 22)$
  d $14 - 21 \div 3$
  e $3 + 2 \times 8$
  f $29 - 2 \times 10$
  g $3 \times 4 - 2$
  h $(20 + 5) \times 3$
  i $5 \times 4 + 30 \div 10$
  j $24 \div 8 \times 6 - 5$
  k $25 + 14 \div 2 - 20$
  l $3 \times 3 - 4 \times 2$
  m $15 \div 3 - 3 - 2$
  n $7 - 24 \div 6$
  o $8 \times 3 \div 4$
  p $5 + 36 \div 6$
  q $54 - 3 \times 8$
  r $40 - 10 \times 3$

7 Work carefully and follow the rules to do these calculations.

a $4 + 8 \times 2 - 25 \div 5 \times 2 + 4$ b $100 - 10 \times 5 - 30 \div 3 \div 5 + 12$

c $3 + 21 \div 7 + 4$ d $3 + 18 \div 9 + 7$

e $18 + 5 \times 1 - 6 \times 3$ f $12 - 2 - 5 \times 2 + 8$

g $18 - (13 - 8 - 2) \times 3$ h $(6 + 4 \times 5) - 2 \times 4$

8 Check Marc's homework and correct any answers he got wrong.

a $3 + 2 \times 8 = 19$ b $39 - 2 \times 6 = 222$

c $3 \times 4 - 2 = 10$ d $(20 + 5) \times 3 = 75$

e $(3 \times 2) + 7 \times 3 = 39$ f $5 \times 4 + 30 \div 10 - 3 = 2$

Challenge

9 Use the four digits of the current year together with the operation signs and brackets if you need them to make up 10 different calculations. Each calculation must have at least two operations.

Exchange your work with another student and do each other's calculations. Check each other's answers when you have finished.

What did you learn?

1 Calculate.

a $4 \times (13 - 5)$

b $5 \times (6 - 3)$

c $(6 \times 8) - (10 + 10)$

d $(6 + 7) - (12 \div 4)$

2 A teacher gave her class this calculation:

$2 \times (8 - 3) - 2 \times 2$

Kamaya got 22. Toniqua got 6.

Sam got 16. Leroy got 12.

Who was correct? Try to work out what the others did wrong.

Topic 6 Review

Key ideas and concepts

To complete this diagram, go back through the units in this topic. For each one, write down the main things you learnt.

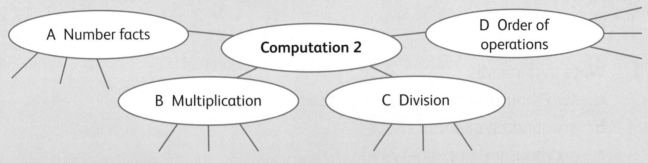

Think, talk, write …

Make up five division story problems.

Use one of the terms from the box in each problem.

equal share	equal groups	quotient	left over	how many

Quick check

1   Calculate mentally. Write the answers only.

  a 4×8 b 8×6 c 9×3 d 6×6 e 7×6
  f $36 \div 6$ g $32 \div 8$ h $25 \div 5$ i $54 \div 6$ j $42 \div 6$

2   Multiply. Show your working.

  a 3×124 b 129×8 c 24×12 d 23×123

3   Divide. Show your working.

  a $78 \div 6$ b $235 \div 2$ c $84 \div 5$ d $564 \div 3$ e $73 \div 4$

4   Calculate.

  a $20 \times 3 + 4$ b $4 + 8 \div 2$ c $5 + 6 \times 2 - 2$ d $15 - 3 \times (12 - 8)$

Problem-solving

5   256 tickets that cost $23.00 each are sold for a cricket final. What is the total cost?

6   How many lengths of 9 m can you cut from a 200 m long piece of rope?

7   A taxi may carry a maximum of 4 passengers. What is the fewest number of taxis needed to transport 34 people?

Test yourself (1)

Explain

Complete this test to check that you have understood and can manage the work covered in Topics 1 to 6.
Revise any sections that you find difficult.

1   Write in numerals.

    a   two thousand four hundred and sixty

    b   six hundred and four

    c   six thousand and ninety-nine

2   Write in words.

    a   809   **b**   717   **c**   8 670   **d**   9 999

3   Copy each statement. Replace the * symbol with a digit to make the statement true.

    a $4\,230 < 4\,{*}30$ **b** $2\,899 < 2\,{*}01$ **c** $8\,3{*}9 < 8\,35{*}$

4   What is the value of the 6 in each number?

    a   3 675   **b**   6 894   **c**   2 006   **d**   3 465

5   Round off each number in the calculations and estimate the answer.

    a $545 + 899$ **b** $469 + 675$ **c** $919 + 564$

    d $473 + 709$ **e** $954 - 781$ **f** $1\,207 - 999$

6   List all the factors of each number.

    a   12   **b**   39   **c**   98

7   Which even numbers come between 187 and 203?

8   Which of these numbers are prime numbers?

52	23	49	67	91	53

9   What is the highest common factor of 32 and 60?

10   Use the word 'factors' or the word 'multiples' to describe each of these sets of numbers.

    a   21, 28, 35, 42, 49   **b**   10, 20, 30, 40, 50   **c**   1, 13

    d   1, 2, 3, 4, 6, 12   **e**   1, 3, 9

11   What is the lowest common multiple of 2, 3 and 7?

12 Calculate. Estimate first and show your working.

    a   67 + 23 + 391   **b**   127 + 49 + 761   **c**   362 + 1 947

    d   783 – 121   **e**   486 – 137   **f**   7 841 – 1 952

13 Draw these shapes. Write one property of each shape below it.

    a   rectangle   **b**   square   **c**   trapezium

14 Which of these triangles are right-angled?

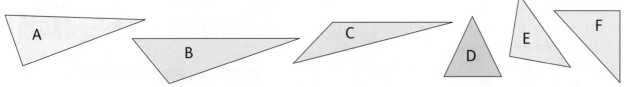

15 The table shows the number of people who bought tickets for a cricket match.

Day 1	Day 2	Day 3
1 463	2 805	2 593

    a   How many tickets in total were sold on the first two days?

    b   How many more tickets were sold on day 3 than on day 1?

    c   On day 4, the home team was doing really well and three times as many tickets were sold as on day 1. How many tickets were sold on day 4?

    d   On day 5, ticket sales were double the sales of day 3. How many tickets were sold on day 5?

    e   The cricket association donated $4.00 to charity for every ticket sold on day 2. How much money did the charity receive?

16 What is 325 more than the difference between 201 and 65?

17 Calculate.

    a   12 + 48 ÷ 6   **b**   (12 + 48) ÷ 6   **c**   13 + 9 – 8 ÷ 4

18 A calculator display shows 993 at the end of a calculation. If the starting number was 8 993, what single operation was done to get that answer?

19 When Sonia rounded a number to the nearest 10 her answer was 80. What possible numbers could she have started with?

20 What number with digits that add up to 14, is <120, >70, odd and a multiple of 5?

21 What number is exactly halfway between 37 and 51?

Teaching notes

Metric units

* Units in the metric system work in multiples of 10 and allow you to write measurements in equivalent ways. For example, 1 cm is equivalent to 10 mm.

* It is useful to have a chart or a number line in the classroom that shows equivalent measures, so that students learn to recognise 'bigger' and 'smaller' units of length, mass and capacity.

Length

* Students need a lot of practice before they are able to estimate and measure accurately. Help students to improve their estimating and measuring skills by working with concrete objects and real-life situations.

* Young students may struggle with measuring length because they
  - have not fully grasped how long a centimetre or a metre really is
  - may not fully understand where to start their measurements
  - don't realise that they should not leave gaps when measuring.

Mass

* You may want to explain to students that we often talk about the 'weight' of something when we really mean the 'mass'. We usually say, 'I weigh 40 kilograms', not 'I have a mass of 40 kilograms'. While it is true to say, 'I weigh 40 kilograms', mass and weight are not exactly the same.

* As with length, it is important to give students lots of practical measuring tasks, so that they can develop a sense of mass and also learn to read the scale on various measuring instruments.

Capacity

* Students have already learnt about capacity and estimating and measuring liquids in litres. Remind them that the capacity of a container is a measure of how much it can hold.

* Continue to let them work with concrete objects when estimating and measuring capacity.

A

Long ago, people used their arm lengths to measure cloth. What problems could there be with this method of measuring? How do you use a measuring tape like the one in the photo to measure a length of cloth?

B

What is this boy measuring? How is he measuring this? Why do you think he is measuring this?

C

How much water do you think this sink can hold – 1 litre or 10 litres? Which container can hold more water – the sink or the green plastic container? How did you decide?

Think, talk and write

A **Length** *(pages 62–63)*

1 Which of these units of measurement will you NOT find on your ruler? Why?

millimetres	kilometres	centimetres
metres		handspan

2 What is an armspan? What would you use it for?

3 Would it be better to measure these in metres or kilometres? Why?
   a the distance between two towns
   b the distance between your house and your neighbour's house

B **Mass** *(pages 64–66)*

1 What is your mass? (How much do you weigh?)

2 What units of mass would you expect to see on each of these packets of food?
   a a bag of sugar
   b a bar of chocolate
   c a sack of potatoes

C **Capacity** *(pages 67–68)*

1 How do you know how much liquid a bottle can hold?

2 You buy a can of soda that says 340 mℓ on the outside. Is this the measurement of the size of the can or of the amount of liquid in the can?

A Length

Last year you measured **length** in **metres** and **centimetres**.

In the metric system we use metres to measure longer lengths. The abbreviation for metres is m.

Your armspan is a **distance** of about 1 metre.

This is one centimetre. ⊢——⊣
The abbreviation for centimetres is cm.
There are 100 centimetres in a metre.
100 cm = 1 m

Some lengths are shorter than 1 centimetre. We can measure these in smaller units called **millimetres**.
The abbreviation for millimetres is mm.

Look at the ruler. You can see that each centimetre is divided into ten smaller units.

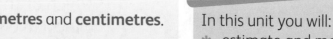

Each of these smaller units is 1 millimetre. There are 10 millimetres in 1 centimetre.
10 mm = 1 cm

A millimetre is about the same as the thickness of a coin.

When you measure length, you write the measurement using numbers and units. For example:

Length of a pencil = 12 cm

Thickness of a mobile phone = 9 mm

Height of the door = 2 m

Sometimes you may need two different units for one measurement. The **width** of a classroom may be 4 m and 55 cm and the length of a mobile phone may be 13 cm and 5 mm.

Maths ideas

In this unit you will:
* estimate and measure lengths in metres, centimetres and millimetres
* compare the length and height of different objects using different units of measurement.

Key words

length	distance
metres	millimetres
centimetres	height
metric system	width

1   Would you measure the following in metres, centimetres or millimetres?
   a   the length of your book
   b   the width of your little finger
   c   the height of the room
   d   the distance to the school gate
   e   the length of an ant

2 **a** Choose five lengths in your classroom that are about 1 metre long. Estimate and then measure each length in centimetres. Record the results in a table.

   b Check the measurements that your partner has made. Are they accurate?

3 You will need a ruler marked in centimetres and millimetres to measure the length of each caterpillar.

   a Measure each length in centimetres and record it.

   b Measure each length again, this time in millimetres, and record it next to the previous measurement.

   c Compare the centimetre and millimetre measurements. What do you notice?

A

B

C

D

E

4 Which measurement in each pair is longer:

   a 303 cm or 518 cm?   **b** 1 600 m or 1 060 m?   **c** 208 mm or 280 mm?

   d 35 mm or 8 cm?   **e** 1 m or 340 cm?   **f** 2 100 mm or 21 cm?

5 Arrange these measurements in order from the shortest to the longest.

   a the distance from your home to your school

   b the length of a broomstick   **c** the height of the school roof

   d the thickness of a hair   **e** the length of a caterpillar

6 Draw and label these line segments in your book. Swap books with a partner and check that he or she measured correctly.

   a 12 cm   **b** 12 mm   **c** 4 cm and 5 mm   **d** 65 mm

Challenge

7 Roger's stride length is 80 cm when he is running. How many strides will he need to take to run 50 m?

What did you learn?

1 Estimate the length of each line in centimetres.

   A ⊢————————⊣   B ⊢————————————⊣   C ⊢————⊣

2 Measure the lines in centimetres (and millimetres, if you need to) and record the measurements.

B Mass

Explain

Mass is a measure of how heavy or light things are. We often use the word '**weight**' when we talk about how heavy or light things are, but the correct word for this in mathematics is 'mass'.

Kilograms (kg) and **grams** (g) are metric units of mass.

Mass can be measured in grams (g) and kilograms (kg) using a balance **scale** or a digital scale.

A kilogram is heavier than a gram, so it is used to measure larger masses.

1 kilogram (kg) = 1 000 grams (g)

These items are all lighter than 1 gram.

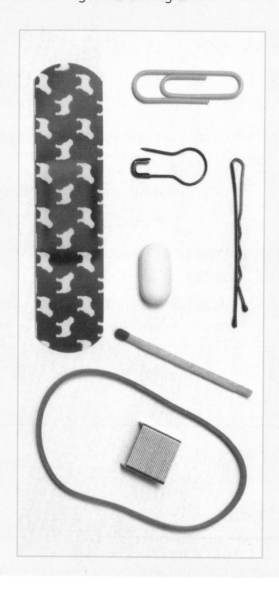

Maths ideas

In this unit you will:
* estimate and measure mass in kilograms and grams.
* compare the mass of different objects using different units of measurements.

Key words

weight

grams

scale

metric system

mass

length

capacity

kilograms

Did you know?

The prefix *kilo-* is used in the **metric system** for **mass**, **length** and **capacity**. Kilo- comes from a Greek word meaning 'a thousand'.

Think and talk

Micah has only three kinds of mass pieces: 1 kg, 3 kg and 9 kg. He only has one of each. He says he can measure all the whole number masses from 1 to 10 **kilograms** using only these three mass pieces. How could he do this?

1   Work in pairs. Use a balance scale to measure the mass of objects in your classroom. Place an object or several objects with the following mass on one side of the scale and then find other objects with the same mass to put on the other side.

   a   10 g
   b   25 g
   c   1 kg

2   What is the mass of each object? Choose the best estimate.

   a

   50 g   500 g   5 kg

   b

   1 g   100 g   1 kg

   c

   50 g   5 kg   50 kg

   d

   2 g   200 g   2 kg

   e

   4 g   4 kg   400 g

3   Martin weighed himself and found his mass to be 37 kg and 500 grams. He is going on a plane trip and can only carry a bag of 7 kilograms. His bag was too bulky to fit on the bathroom scale, so he held it and weighed himself with the bag. The scale showed 45 kg and 200 g. Is his bag light enough to carry on board?

4   Marcia decided to bake a cake. This is the list of ingredients for the cake.

Marcia's mom had these full packets of flour, sugar and cocoa powder at home.

   a   How much was left in each packet after Marcia made the cake?

   b   What would Marcia have to buy if she wanted to make two cakes?

| 350 g flour |
| 4 eggs |
| 150 g sugar |
| 70 g cocoa powder |

What did you learn?

Are these statements true or false?

1   4 kg is heavier than 3 000 g

2    $100 \times 1\ g = 1\ kg$

3   A gram is a metric unit of mass.

4   A mouse has a mass of about 35 kg.

C Capacity

The amount of liquid that a container can hold is called its **capacity**.

We can use **litres** (ℓ) and smaller units called **millilitres** (mℓ) to measure how much liquid a container holds.

1 litre (1 ℓ) = 1 000 millilitres (mℓ)

MILK 1 LITRE

MILK ½ ℓ

MILK 500 mℓ

In this unit you will:
* estimate and measure capacity in litres and millilitres
* compare the capacity of different containers using different units of measurement.

capacity

litres

millilitres

1   Work in groups. Collect ten different containers.

   a   Estimate the capacity of each container. Then arrange the containers in order from the smallest capacity to the greatest capacity.

   b   Use water, a measuring spoon and a measuring jug to measure the capacity of each container.

2   Which unit would you use to measure the capacity of each of these containers: litres or millilitres?

   a   a tea cup

   b   a bottle that holds liquid soap

   c   a can of petrol

   d   the lid of a glass bottle

3   Write the following capacities in millilitres.

   a   1 litre   b   3 litres   c   ½ litre   d   ¼ litre

4   a   What is the capacity of this jug?

   b   How much juice is in the jug at the moment?

   c   How much more juice is needed to make 500 mℓ?

5   1 teaspoon = 5 mℓ and 1 cup = 250 mℓ

   Use the given information to work out the capacity of these containers.

   a   2 cups   **b**   5 teaspoons   **c**   11 cups   **d**   8 teaspoons

6   **a**   How many millilitres of water are in each jug?

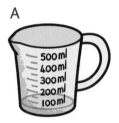 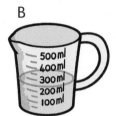 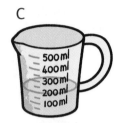 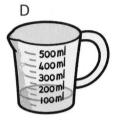 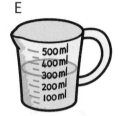

   b   Write the amounts in order from the least to the most.

   c   Which two jugs contain 450 mℓ between them?

Problem-solving

7   Mr Ashburton is given a 125 mℓ bottle of medicine and told to take two 5 mℓ spoonfuls twice per day. How many days will the bottle of medicine last?

8   There is a leaky tap in the kitchen. The tap leaks 1.5 ℓ of water every hour. You need to put a bucket under the tap to collect the water so that you do not waste it. The bucket you have has a capacity of 12 ℓ. After how many hours will you need to empty the bucket?

9   Imagine you have three buckets with the following capacities: 7 ℓ, 5 ℓ and 3 ℓ. You need to measure out 4 ℓ of water. Work out how you can use the buckets to do this.

What did you learn?

Which unit of measurement would you probably find on the label of each container?

1   a small carton of juice

2   a large bottle of tomato ketchup

3   a bottle of cough mixture

4   a large bottle of water

Topic 7 Review

Key ideas and concepts

Answer these questions to summarise what you learnt in this topic.

1   What units can you use to measure shorter lengths? What is the relationship between them?
2   What do we measure in grams? How many grams are there in one kilogram?
3   What does the word 'capacity' mean?
4   Which is the larger unit: the millilitre or the litre?

Think, talk, write …

What is the mathematical word for:

1   measuring how long something is
2   the amount of liquid that a container can hold
3   a very small unit of capacity
4   the heaviness or lightness of an object

Quick check

1   Measure the length of this line segment using two different units of measurement. ├─────────────┤

2   a   How many grams are there in 2 kilograms?
    b   How many kilograms is equivalent to 3 000 grams?

3   Would you measure these masses in grams or kilograms?
    a   a bar of chocolate
    b   a large sack of oranges
    c   your own body mass

4   Write these capacities from the greatest to the smallest.
    1 teaspoon   1 ℓ   250 mℓ   4 000 mℓ   1 cup $\frac{1}{2}$ litre

5   Jessica's dog eats two tins of food every day. Each tin has a mass of 180 grams. How many grams of food does the dog eat:
    a   in a day?   b   in a week?

A

Which do you think are more popular – cones or stick ice-creams? How could you find out? How can knowing this help an ice-cream vendor?

Teaching notes

Methods of collecting data

* Students used observations and simple interviews to collect data last year. This year they will also work with simple questionnaires.
* Observation involves looking and often counting.
* An interview involves asking questions to collect data.
* A questionnaire is a form that is used to collect and record data.

Tally and frequency tables

* Tallies are little strokes, like this: ////. When you get to five, you draw a line across the previous four strokes like this: ####. This makes groups of five, which can easily be totalled by skip counting.
* A frequency table is a way of organising data to show the frequency of data items (how many there are) in each category or group.
* Some tables combine tally marks and frequencies.

B

| Plain | ⧸⧸⧸⧸ ⧸⧸ |
| Patterned | ⧸⧸⧸⧸ ⧸ |

Tammy wants to know how many plain and how many patterned beads there are in the tray. She uses the table below to help her. What type of table is this? How can it help you organise information?

(pages 72–74)

A Collecting and organising data

(pages 72–74)

1 A teacher wants to find out what her students' favourite subjects are.
   a How could she find this data?
   b How would you organise data about students' favourite subjects? Why?

2 The teacher also wants to know how well students perform in mathematics tests. How could she find this out?

B Collect and organise your own data

(pages 75–76)

Mike asked 18 of his friends how they get to school in the morning. These are his results.

car	car	walk	bus	walk	walk
bus	walk	bike	car	bus	walk
walk	bus	bus	car	bike	walk

1 Why is this data difficult to work with?

2 Draw up a tally chart to organise the data.

3 How does the chart make the data easier to work with?

A Collecting and organising data

Explain

Data is information you can **collect**, **record** and organise. You can collect data by observation and doing interviews.

Observation – looking, counting, measuring and making notes. For example, you can look around your class and count the number of children who wear spectacles.

Interviewing – asking questions orally and recording the answers. For example, you can ask people how many times a day they check their phone for messages.

You can also collect data using a **questionnaire**. This is a form with questions that people fill in and give back to you. They can do this in writing or type the answers on a computer. Once you get the questionnaires back, you have to sort and organise the data yourself.

Look at this example of a questionnaire.

School lunch survey
Please circle the correct information.
Gender: Male   Female
Age: 4 5 6 7 8 9 10
1. Do you pack a lunch for school?   Yes   No
2. If yes, what sort of container/packaging do you use?
    Wax wrap Plastic bag Cling wrap Tinfoil
    Lunchbox
    Ice-cream container Tin Other:_____
3. If you buy lunch, where do you buy it?
    Tuckshop Café Canteen Vendor Local market

Maths ideas

In this unit you will:
* learn about three different methods of collecting data and what type of data is suited to each method
* collect your own data using observation, interviews and questionnaires
* use tallies and frequency tables to record and organise data.

Key words

data
collect
record
observation
interviewing
questionnaire
survey
tallies
tally chart
frequency table

Explain

Tallies allow you to keep a count without writing numbers. Think about trying to keep track of how many men, women and children pass by during a carnival parade. A **tally chart** like this one allows you to record each person quickly.

Men	//// //// //// //// //// //// ////
Women	//// //// //// //// //// ////
Children	//// //// //// //// //// //// //// /

When you have finished tallying, you can add up the marks by counting in fives to get a total.

Men	34
Women	30
Children	36

A table with totals is called a **frequency table**.
Some tables have both tallies and frequencies.

1   Which method do you think is most suitable for collecting the following data?
   a   how satisfied patients are with waiting times at a clinic
   b   the most popular make of car sold on your island
   c   the amount of time students spend watching TV during the week
   d   how many children in your school wear spectacles
   e   how much money people are prepared to spend on health care
   f   national census information about every household on your island

2   List at least five situations in which you could carry out an observation to collect data.

73

3   Maria and Jose carried out an observation to find out which of the two entrances at their school was used by most students.

   a   How do you think they did this?

   b   How do you think they recorded the information?

   c   How could you find this data for your own school?

4   Maria and Jose used tallies to keep track of how many students passed through each entrance. These are their results.

Front Gate – St Michael's Road
ＩＨＨ ＨＨ ＨＨ ＨＨ ＨＨ ＨＨ
ＨＨ ＨＨ ＨＨ ＨＨ ＨＨ ＨＨ
ＨＨ ＨＨ ＨＨ ＨＨ ＨＨ ＨＨ
ＨＨ ＨＨ ＨＨ ＩＩＩ

Back Gate – Garvey Avenue
ＨＨ ＨＨ ＨＨ ＨＨ ＨＨ
ＨＨ ＨＨ ＨＨ ＨＨ ＨＨ
ＨＨ ＨＨ ＨＨ ＨＨ ＨＨ
ＨＨ ＨＨ ＨＨ ＩＩＩＩ

   a   How does this system save time and space?

   b   How many students passed through each entrance?

What did you learn?

1   List at least five situations in which you could interview students at your school to collect data.

2   What is the difference between an interview and a questionnaire?

B Collect and organise your own data

Explain

When you do an **interview** or use a **questionnaire** to **collect data**, you ask questions to find the information you want. The questions you ask should be clear and easy to understand. People can answer the questions by checking boxes, by marking a choice from a number of given answers, or by writing their own answers.

I am a boy ✓ girl ☐

Age in years: 7　⑧　9　10　11　12

Favourite sport: basketball

Maths ideas

In this unit you will:
* design your own questionnaires to collect data
* carry out some investigations to collect and organise data.

Key words

interview

questionnaire

collect

data

organise

1　Jared wanted to find out what other students think about litter on the school grounds. He made this questionnaire to collect data.

a　What questions do you think Jared asked about litter before he wrote up this questionnaire?

b　Is it a good questionnaire or not? Give reasons for your answer.

Questionnaire: Litter at school

Please circle your level.

1　　2　　3　　4　　5　　6

Please tick the correct box.

1　Is litter a problem at our school?

　　Yes ☐　　　No ☐　　　I'm not sure ☐

2　Do we need more bins at our school?

　　Yes ☐　　　No ☐　　　I'm not sure ☐

3　Are you willing to take part in an anti-litter campaign?

　　Yes ☐　　　No ☐　　　I'm not sure ☐

4　Which of these solutions would work best to reduce litter at our school?

　　(Tick up to three boxes.)

　　More bins ☐　　　　　　　　Litter monitors ☐

　　Punishment for people who litter ☐　More recycling projects ☐

　　Rewards for picking up litter ☐　　Close the tuck shop ☐

5　Do you have any suggestions for reducing litter at school? Write them here.

　　_____

　　_____

　　_____

2   Jamira wants to find out what type of music people prefer to listen to. Her question is: 'Do you like rap or pop music?' Is this a suitable question for a questionnaire? If not, how can it be improved?

Investigate

3   Work in groups of four.

Design a questionnaire to find the following information about your classmates.
* age   * gender (boy or girl)
* number of brothers and sisters   * favourite subject
* how they travel to school   * whether they wear spectacles or not

   a   How could you record and **organise** the data you collect?
   b   Carry out your investigation. At least 12 classmates should answer your questionnaire.
   c   Record and organise the data.
   d   Then write a few sentences to summarise what you found out.

4   Work in a group again.
     Consider these questions.

   a   For each question, draw up a simple questionnaire that you could use to collect this data.
   b   Choose one of the questions. Use your questionnaire to collect data from at least 20 students in your school.
   c   Organise your data using tables or graphs or both.
   d   Summarise what you found out.

How many people live in each household?

Do you and your family recycle or not?

Do you grow any of your own food?

How many times a week do you use public transport?

What did you learn?

Write a set of instructions to help someone draw up a good questionnaire to find out how many hours people spend on different activities each day, such as homework, cooking, sleeping, watching TV, chores, and so on.

Topic 8 Review

Key ideas and concepts

1   Copy this table into your book and write short notes in each space to summarise what you know about collecting and organising data.

Method of collection	What does this mean?	What is this method useful for?
Observation		
Interview		
Questionnaire		

2   Write a short definition of each of these key terms.

a   tally chart   b   frequency table

Think, talk, write …

For each person, give an example of data they might need and suggest how they could collect and organise the data.

1 . the head of children's programming at a radio station

2   a vendor at a primary school

3   the editor of a magazine for young people

4   a nurse at a local clinic

5   a school principal at the start of a new school year

Quick check

Look at the words in the box.

observation	interview	recording	measuring	questionnaire

a   Draw up a tally chart and use it to count how many times the vowels (a, e, i, o and u) appear in the words.

b   Complete the table by adding the frequency for each vowel.

c   Which vowel appears 7 times?

Teaching notes

Fractions

* A fraction is a part of a whole. In mathematics, the whole is divided into equal parts to make fractions.
* The 'whole' can be an object, for example, a shape or area. Or it can be a number of individual objects grouped as a set.
* Common fractions are written using two numbers and a fraction line, for example, $\frac{1}{3}$ and $\frac{7}{10}$.
* The number below the line is called the denominator. It tells you into how many equal parts the whole is divided.
* The number above the line is called the numerator. It tells you how many equal parts you are dealing with.

Mixed numbers and improper fractions

* When the fractional parts combine to make more than one whole, they can be written as mixed numbers or improper fractions. For example, if you have three half bars of chocolate, you can put them together to make 1 whole and 1 half, which can be written as the mixed number $1\frac{1}{2}$.
* Three halves can also be written as $\frac{3}{2}$. When the numerator is greater than the denominator like this, it is called an improper fraction.
* It is important to use concrete examples to introduce these very important concepts.

Equivalent fractions

* A fraction wall shows that the same part of the whole can be expressed using different fractions. For example, one half can also be expressed as two quarters or five tenths.
* Common fractions that represent the same part of the whole are equivalent. This means that $\frac{1}{2} = \frac{2}{4} = \frac{5}{10}$.
* A mixed number and an improper fraction can be equivalent. For example, $\frac{5}{2}$ means 5 halves, which is equivalent to $2\frac{1}{2}$ (two wholes and one half).

A

Zara stuck these stickers onto her pencil case. How many stickers are there? How many stars are there? What fraction of the stickers are stars? What fraction of the stickers are circles? What fraction of the circles are pink?

1234

B

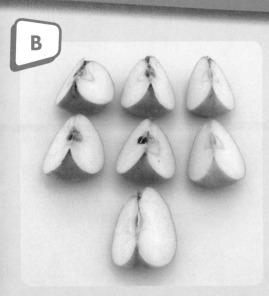

Each of these pieces of apple is $\frac{1}{4}$ of a whole apple. How many quarters are shown here? How could you write this as a fraction with a denominator of 4? If you put the pieces back together, how many apples would you have?

C

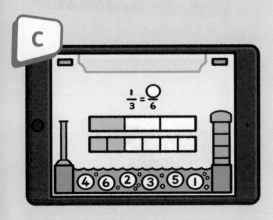

Jeffrey is playing a mathematics game on his computer. Which bubble should he drag to get this question correct? If the next question is $\frac{2}{3} = \frac{\square}{6}$, which bubble should he drag to get it correct?

Think, talk and write

A Revisiting fractions *(pages 80–81)*

What fraction of each shape is coloured?

1

2

3

4

B Mixed numbers and improper fractions *(pages 82–83)*

Eugenia bought some cookies to share with her friends. They were quite big, so she broke them into quarters. At the end of the day there were some quarter cookies left over.

1 How many quarters does she have left?

2 If you combine the pieces, how many cookies are there altogether?

C More equivalent fractions *(page 84)*

Look at these fraction bars.

1 Express the yellow part of each bar as a fraction.

2 What can you say about the four fractions you have written down?

3 Find the bar that has $\frac{6}{9}$ green. Write one other fraction that is equivalent to $\frac{6}{9}$.

A Revisiting fractions

The **whole** circle below has been divided into five **equal parts**. Each part is one fifth, or $\frac{1}{5}$, of the whole circle.

2 out of 5 pieces are red.

$\frac{2}{5}$ are red

numerator

$\frac{2}{5}$

denominator

The numerator of a **fraction** tells you how many parts you are working with.

The denominator tells you how many equal parts are in the whole.

If you look at the diagram, you can see that:

$\frac{2}{5}$ red + $\frac{3}{5}$ white is equal to $\frac{5}{5}$, which is the same as the whole circle.

In mathematics, we say that $\frac{5}{5} = 1$.

Remember, when the numerator and denominator are equal, the fraction represents one whole.

$$\frac{2}{2} = 1 \qquad \frac{3}{3} = 1 \qquad \frac{4}{4} = 1 \qquad \frac{8}{8} = 1 \qquad \frac{12}{12} = 1$$

Maths ideas

In this unit you will:
* revise what you learnt about fractions last year
* use fraction bars to compare and order fractions
* solve problems involving fractions.

Key words

whole	fraction
equal parts	compare
numerator	order
denominator	

Important symbols

> greater than
< smaller than
= equal to

1 Write the fraction that is circled or shaded in each example.

2 Draw a diagram to show each fraction.

    a $\frac{2}{3}$ b $\frac{3}{5}$ c $\frac{5}{6}$ d $\frac{7}{7}$

3 Which of the fractions in Question 1 has a numerator of 5?

4 Which fraction in Question 1 can be written with a denominator of 3?

5 Which of the fractions you drew in Question 2 can be written as 1?

6 Micah colours some shapes blue and yellow. The blue fraction of each shape is given here. What fraction is yellow?

    a $\frac{3}{7}$ b $\frac{2}{9}$ c $\frac{1}{5}$ d $\frac{1}{4}$

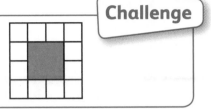

7 The diagram shows a table top with its centre painted red. What fraction of the table top is not painted red?

Challenge

Explain

When fractions have the same denominators, the one with the greater numerator is the greater fraction, as it contains more parts of the whole.

For example, $\frac{1}{10} < \frac{8}{10}$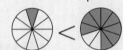

To **compare** two fractions with the same numerator, but different denominators, look at the denominators. The fraction with the bigger denominator is the smaller one because the whole is divided into more parts.

$$\frac{1}{2} > \frac{1}{3} \qquad \frac{1}{3} > \frac{1}{4} \qquad \frac{3}{5} < \frac{3}{4} \qquad \frac{4}{10} < \frac{4}{9}$$

To compare other fractions, you need to think about the fraction wall and how much of the whole you are working with. Look at these examples carefully.

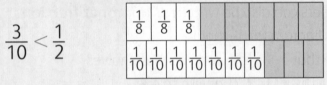

$$\frac{3}{10} < \frac{1}{2} \qquad \qquad \frac{3}{8} < \frac{7}{10}$$

8 Write true or false for each statement.

a $\frac{1}{4} > \frac{1}{2}$ b $\frac{1}{3} > \frac{1}{10}$ c $\frac{3}{5} < \frac{3}{4}$ d $\frac{4}{7} > \frac{4}{9}$

9 Rewrite the statements. Fill in <, = or > to make each one true.

a $\frac{4}{4} \square \frac{3}{5}$ b $\frac{1}{2} \square \frac{4}{5}$ c $\frac{3}{6} \square \frac{1}{2}$ d $\frac{4}{9} \square \frac{5}{8}$

10 Rewrite each set of fractions in **order** from the smallest to the greatest.

a $\frac{1}{10}, \frac{1}{8}, \frac{1}{2}, \frac{1}{3}$ b $\frac{2}{5}, \frac{2}{9}, \frac{2}{10}, \frac{2}{2}$ c $\frac{3}{4}, \frac{2}{5}, \frac{3}{8}, \frac{2}{3}$

What did you learn?

1 Which of these fractions is closest to 0? Give reasons to support your answer.

$$\frac{2}{5} \qquad \frac{1}{2} \qquad \frac{3}{10} \qquad \frac{1}{4}$$

2 How would you prove to someone that $\frac{5}{6}$ is greater than both $\frac{7}{12}$ and $\frac{5}{8}$?

B Mixed numbers and improper fractions

So far you've worked with **fractions** that are smaller than, or equal to, one **whole**. Now you are going to learn about fractions that show amounts greater than one whole.

> **Example**
> Jerome bought two bars of chocolate.
> He ate half of one bar.
> He has one and a half bars left.

We write this in numbers as $1\frac{1}{2}$ and we call it a **mixed number**.

A mixed number is made up of a whole number and a fraction. For example:

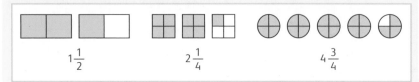

$$1\frac{1}{2} \qquad 2\frac{1}{4} \qquad 4\frac{3}{4}$$

Mixed numbers can also be written as **improper fractions**.
Look at the diagram for $1\frac{1}{2}$ again.
Can you see that $1\frac{1}{2}$ is the same as three halves?
We can write this as a fraction like this: $\frac{3}{2}$
We say three halves.
$1\frac{1}{2} = \frac{3}{2}$ We say these two fractions are **equivalent**.
$2\frac{1}{4}$ is the same as 9 quarters.
We write $\frac{9}{4}$ and say nine quarters.
$4\frac{3}{4}$ is the same as 19 quarters.
We write $\frac{19}{4}$ and say nineteen quarters.
You will often use mixed numbers for measurements and when you tell the time.

Maths ideas

In this unit you will:
* learn about mixed numbers and improper fractions
* write the same fraction as a mixed number and as an improper fraction.

Key words

fractions improper
whole   fractions
mixed   equivalent
  number

> I need $1\frac{3}{4}$ metres of fabric to make a skirt.

> It took me $2\frac{1}{2}$ hours to get home.

> I used $1\frac{1}{4}$ cups of flour to make pancakes.

1   Write the shaded part of each set of shapes as a mixed number and as an improper fraction.

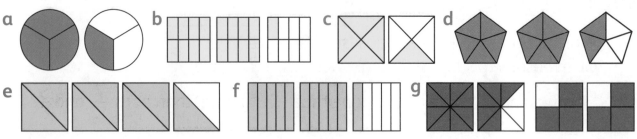

2 How many cookies in each set? Give your answer as a mixed number and as an improper fraction.

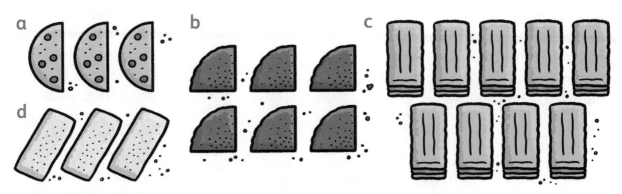

3 Draw your own diagrams to show these mixed numbers.

a $1\frac{1}{2}$ b $2\frac{1}{4}$ c $4\frac{1}{3}$ d $1\frac{3}{4}$ e $1\frac{7}{10}$

4 Write each improper fraction as a mixed number. Draw diagrams if you need to.

a $\frac{5}{2}$ b $\frac{7}{4}$ c $\frac{9}{5}$

d $\frac{8}{6}$ e $\frac{13}{4}$

Think and talk

Why do you think improper fractions are sometimes called 'top heavy'?

5 Write each set of numbers in ascending order.

a $1\frac{1}{2}$ $1\frac{1}{3}$ $1\frac{2}{3}$ $2\frac{1}{4}$ $1\frac{3}{4}$

b $\frac{17}{4}$ $\frac{17}{6}$ $\frac{19}{4}$ $\frac{13}{3}$ $\frac{8}{3}$

Problem-solving

6 Mrs Thomas has 4 metres of rope. She cuts off $\frac{3}{4}$ of a metre. How much rope is left?

7 Which is longer: $1\frac{1}{4}$ years or 18 months?

What did you learn?

Match each improper fraction with its equivalent mixed number. Write the pairs in your book.

$\frac{5}{4}$ $\frac{6}{5}$ $\frac{8}{7}$ $\frac{4}{3}$ $\frac{5}{2}$ $\frac{9}{7}$ $\frac{7}{2}$

$1\frac{1}{3}$ $2\frac{1}{2}$ $1\frac{1}{4}$ $1\frac{2}{7}$ $3\frac{1}{2}$ $1\frac{1}{5}$ $1\frac{1}{7}$

C More equivalent fractions

Equivalent fractions have the same value.
You can find equivalent fractions by multiplying the **numerator** and **denominator** by the same number.

$$\frac{1}{2} \times \frac{2}{2} = \frac{2}{4} \qquad \frac{1}{3} \times \frac{3}{3} = \frac{3}{9} \qquad \frac{1}{2} \times \frac{10}{10} = \frac{10}{20}$$

So fractions like $\frac{2}{2}$, $\frac{3}{3}$ and $\frac{10}{10}$ are all equivalent to 1.

You can also find equivalent fractions by dividing the numerator and denominator by the same number.

$$\frac{10}{20} \div \frac{2}{2} = \frac{5}{10} \qquad \frac{5}{10} \div \frac{5}{5} = \frac{1}{2} \qquad \frac{8}{12} \div \frac{4}{4} = \frac{2}{3}$$

When you do this, we say you are simplifying the fraction.
When you cannot divide the numerator and denominator by any number (except 1), the fraction is in its simplest form. You cannot **simplify** it any further.

1 Write three different equivalent fractions for each fraction given.

    a $\frac{1}{3}$ b $\frac{2}{5}$ c $\frac{3}{4}$ d $\frac{5}{6}$ e $\frac{3}{10}$

2 Simplify each fraction and write it in simplest form.

    a $\frac{6}{12}$ b $\frac{5}{10}$ c $\frac{3}{18}$ d $\frac{14}{21}$ e $\frac{20}{60}$

3 These fractions are equivalent. What is the missing number in each pair?

    a $\frac{4}{8} = \frac{\square}{4}$ b $\frac{6}{10} = \frac{\square}{5}$ c $\frac{3}{5} = \frac{\square}{20}$ d $\frac{\square}{3} = \frac{6}{9}$

In this unit you will:
* use diagrams to revise equivalent fractions
* multiply and divide to make equivalent fractions
* understand how to simplify a fraction.

equivalent   denominator
numerator   simplify

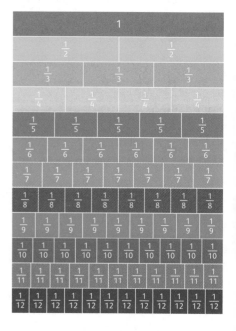

What did you learn?

1 Look at the diagrams.

    a What fraction of cake did each child receive?

    b What can you say about these fractions?

Anne's share   Jamie's share   Tori's share   Gershwin's share

2 How can equivalent fractions help you to order and compare fractions?

Topic 9 Review

Key ideas and concepts

Complete these sentences to summarise what you learnt in this topic.

1 A fraction is _____.
2 The denominator shows _____.
3 The numerator tells you _____.
4 To order and compare fractions you need to _____.
5 A mixed number has a _____.
6 An improper fraction is one where _____.
7 Equivalent fractions have _____.
8 You can find equivalent fractions if you _____.

Think, talk, write …

Fractions and mixed numbers are used in everyday conversations all the time.

1 Complete these sentences by using fractions in a sensible way.

   a I would like half a _____.
   b The trip takes one and three quarters _____.
   c Just over $\frac{2}{3}$ of the _____.

2 Make up three everyday sentences of your own using fractions and mixed numbers. Write them in your maths journal.

Quick check

1 Here are six shapes with blue and yellow parts.

   a Write the fraction of each shape that is blue.
   b Order the fractions from the smallest to the greatest.
   c Write two equivalent fractions for the yellow part of each shape.

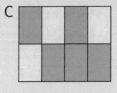
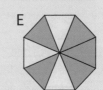
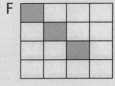

Problem-solving

2 Nicky wants to share five cakes equally among four children. What fraction of the cakes will each child receive?

Teaching notes

Number patterns

* In mathematics, patterns can repeat or grow.
* The pattern red, blue, red, blue, red, blue is a repeating pattern.
* Counting forwards produces a growing pattern. For example: 2, 4, 6, 8, …
* Counting backwards is also considered a growing pattern. For example: 100, 90, 80, …
* To work out missing numbers in a pattern, or to continue a number pattern, you have to identify the pattern and work out whether the numbers are ascending or descending.

Rules for number sequences

* Each number in a sequence is called a term.
* Number sequences can be generated from a rule. The rule tells you what to do to get the next term in a pattern. For example, the rule 'Start at 0 and add 2 each time' would give you the pattern of even numbers: 0, 2, 4, 6, 8 …
* Tables are very useful for listing terms so that you can work out the rule for a particular sequence.

Number sentences with unknowns

* Problems can be expressed in the form of number sentences with an unknown quantity. For example:
* ☐ + 8 = 23
* The ☐ in the number sentence represents the unknown value.
* Writing a number sentence with one or more unknown quantities is a useful problem-solving strategy. At this level, the unknown quantities can be shown using empty boxes or other shapes. Later on, students will use letters (such as x) to represent unknowns.

A

Glen made this pattern on a pegboard. What will the next shape in the pattern look like? How many pegs will he need to make the next pattern? How did you work this out?

B

Mr Julien likes to make chicken curry. He uses three onions for every $\frac{1}{2}$ kg of chicken. If he has $2\frac{1}{2}$ kg of chicken, how many onions will he need?

C

Sharon used 28 beads to make a bangle. She has these beads left over. How many did she have to start with? Can you write a number sentence to show how you worked this out?

Think, talk and write

A   Describing patterns *(pages 88–89)*

Here are two number patterns.

Pattern A: 1, 5, 9, 13 …

Pattern B: 5, 10, 20, 40 …

1   Discuss how each pattern is made.

2   Work out the next three numbers in each pattern.

3   Try to write a rule for making each pattern.

B   Investigating patterns and rules *(pages 90–91)*

One goat has 4 legs, two goats have 8 legs and three goats have 12 legs. You can show this in a table.

Number of goats	1	2	3	4
Number of legs	4	8	12	?

1   How many legs will four goats have?

2   What number pattern is made by the number of legs?

3   How many legs will there be if there are 10 goats?

4   Explain how you can work out the number of legs without counting in fours.

C   Number sentences *(page 92)*

1   Work out the missing number in each number sentence.

   a   4 + ☐ = 8   b   12 − 8 = ☐

   c   3 × ☐ = 12   d   20 ÷ 4 = ☐

2   Tell your partner how you worked out each answer.

A Describing patterns

A **pattern** is an ordered set of objects or numbers.

Numbers in a pattern can be **ordered** in a specific way. Each number in the pattern is called a **term**.

The order of the terms in a pattern helps you to work out what comes next in the pattern.

20, 40, 60, 80, ____, ____, ____

This pattern is in **ascending** order.

The numbers increase by 20 each time.

The next three terms are 100, 120 and 140.

100, 95, 90, 85, ____, ____, ____

This pattern is in **descending** order.

The numbers decrease by 5 each time.

The next three terms are 80, 75 and 70.

You can describe number patterns using the **rule** that is followed to make them.

The rule for the first pattern could be: start with 20; add 20 to get the next term.

The rule for the second pattern could be: start at 100; then **skip count** back in 5s to get the terms in the pattern.

Maths ideas

In this unit you will:
* learn more about number patterns
* find the next numbers in a pattern
* describe the rules for number patterns in words
* follow rules to make number patterns.

Key words

pattern

ordered

term

ascending

descending

rule

skip count

1   Write the next three numbers in each pattern.

   a   142, 144, 146, 148, ____, ____, ____

   b   7, 14, 21, ____, ____, ____

   c   16, 24, 32, 40, ____, ____, ____

   d $\frac{1}{4}, \frac{1}{2}, \frac{3}{4}, 1, 1\frac{1}{4}$, ____, ____, ____

   e   850, 825, 800, ____, ____, ____

2   Read the rule. Write the first five terms of each pattern.

   a   Start at 70 and count back in tens.

   b   Start at 1 450 and count forwards in 100s.

   c   Start at 10 and subtract $\frac{1}{2}$ to find the next term.

   d   Start at 479. Increase each term by 10.

   e   Start at 3. Double the number to find the next term.

3   Write a rule to describe each pattern.

    a   14, 21, 28, 35, …   b   30, 50, 70, 90, …   c $9\frac{1}{2}$, 9, $8\frac{1}{2}$, 8, …

    d   80, 40, 20, 10, …   e   2, 4, 8, 16, 32, …

4   Work out the missing numbers in each pattern.

    a   36, 32, ___ , 24, ___ , 16   b   ___ , ___ , 46, 41, 36, ___

    c   1, 2, ___, 8, ___, 32   d   750, ___, ___, 675, 650, 625

    e   2 400, 1 200, ___, 300, ___

Problem-solving

5   Leshawn made these patterns using matches.

Pattern A     **Pattern B**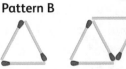

Term 1   Term 2   Term 3   Term 1   Term 2   Term 3

Answer these questions for each pattern.

    a   Draw the next two shapes that Leshawn would make.

    b   Without drawing the next shape, write down the number of matches Leshawn would need to make it.

    c   The first term in each pattern is the same. Why are the number patterns different?

Investigate

6   Use matches, coins or other small objects to make your own pattern.

    a   Draw the first four terms in your pattern.

    b   Swap with your partner. Describe your partner's pattern.

    c   Write a number pattern to match your partner's shape pattern.

What did you learn?

For each number pattern, write the next three terms and write a rule to describe the pattern.

1   2, 8, 14, 20, 26, ___, ___, ___   2   640, 320, 160, ___, ___, ___

3   72, 66, 60, 54, ___, ___, ___   4   5, 9, 13, 17, ___, ___, ___

B Investigating patterns and rules

Explain

When you have a **pattern**, you can usually find the next few **terms** by counting on or back, but it is more difficult to find the tenth term, or the hundredth term by simply looking at the pattern.

A **table** can help you find a **rule** for the pattern and allow you to work out the number of any term in the pattern.

Example

Sondra made this pattern using matches.

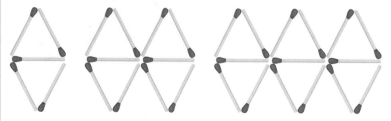

How many matches will she need to make:

a   the next shape?

b   the tenth shape?

Start with a table.

Term number	1	2	3	4	10
Number of matches	5	10	15	?	?

You can see that the pattern involves counting in fives. This means you can look at the five times table.

$1 \times 5 = 5$ This is the first term.

$2 \times 5 = 10$ This is the second term.

$3 \times 5 = 15$ This is the third term.

The rule is: 'multiply the term number by 5'.

The next shape is the fourth term.

$4 \times 5 = 20$ Sondra needs 20 matches.

The tenth shape is the tenth term.

$10 \times 5 = 50$

Sondra needs 50 matches to make the tenth shape in the pattern.

Maths ideas

In this unit you will:
* investigate different patterns to see how they work
* represent shape patterns as number patterns
* use tables to help you work out the rule for a pattern.

Key words

pattern

terms

table

rule

Investigate

* Count in 6s to 100.
* List the numbers you count.
* Are there any odd numbers? How can you tell?
* Add the digits of each number together. What pattern do you get?

1 Look at Jayson's pattern and table.

Term number	1	2	3	4	5
Number of sides	6	12	18	24	?

a What is the rule for finding the number of sides in a term?

b How many sides will there be in the 5th term?

c How many sides will the 100th term have?

2 Rachel made this pattern with toothpicks.

Term number	1	2	3	4	?
Number of toothpicks	5	8	11	?	32

a What is the rule for making each shape?

b How many toothpicks will she need for the 4th shape?

c Which shape in the pattern needs 32 toothpicks?

d How many toothpicks will she need to make the 20th shape in the pattern?

Problem-solving

3 Nikita and Shireen have both made shape patterns using toothpicks. They have each made a table to show the number of toothpicks they used for each term. Use their tables and draw a shape pattern to match each table.

Nikita's table

Term number	1	2	3	4
Number of toothpicks	4	7	10	13

Shireen's table

Term number	1	2	3	4
Number of toothpicks	5	9	13	17

4 Here is another set of numbers.

a List the numbers you would count if you counted in 7s.

b List the numbers you would count if you counted in 8s.

c Compare the sets. What patterns do you notice?

28	32	40
70	72	48
63	49	56
45	54	

What did you learn?

What are the missing numbers in each counting pattern?

1 ____, ____, 18, 24, 30, ____

2 ____, 16, ____, 32, ____, ____

C Number sentences

Explain

When you have to **solve** word problems you can draw a diagram and write a **number sentence** with an empty shape to represent the value you need to work out. When you have worked out the missing value, you have solved the number sentence.

Maths ideas

In this unit you will:
* write number sentences to represent word problems
* solve number sentences to find the unknown values.

Key words

solve
number sentence
unknown

Example

Joe, Richard and James have collected 73 cans for recycling. If Joe collected 22 cans and James collected 37 cans, how many did Richard collect?

Draw a bar diagram.

Write a number sentence.

Use ☐ to represent the **unknown** number.

$22 + ☐ + 37 = 73$

Solve the number sentence.

$22 + 37 = 59$

$73 - 59 = 14$

73 cans

Joe	Richard	James
22	?	37

$-6 \quad -3 \quad -50$
$14 \quad 20 \quad 23 \quad 73 \quad 59$

Write an answer statement: Richard collected 14 cans.

Remember:
Always draw a diagram and write a number sentence to represent each problem before you solve it.

1   Samuel was playing computer games. He scored 470 in the first game, 542 in the second game and only 319 in the third game. How many points did he score in all?

2   Two farmers brought 2 945 coconuts to market. If one farmer brought 1 586 coconuts, how many did the other bring?

3   A tour guide had 4 827 tourists in June and 6 277 tourists in July. How many more tourists did she have in July than in June?

What did you learn?

1   Salma ran 3 750 m on Monday, 4 980 m on Tuesday and 5 409 m on Wednesday. How far did she run altogether?

2   A school store has 3 274 blue and red crayons. If 1 764 are blue, how many are red?

Topic 10 Review

Key ideas and concepts

Match each word on the left to its correct meaning on the right.

Sequence	Going down in order, decreasing
Rule	Going up in order, increasing
Ascending	Instruction for generating a pattern
Skip count	An ordered set of numbers that follows a pattern
Descending	A calculation with numbers and an equals sign
Number sentence	Work out the answer
Solve	Count in groups

Think, talk, write …

Walk around your local area and either take photographs of, or draw, five different patterns that you can see.
Show your patterns to your group and try to describe them in words.

Quick check

1   Write the next three terms in each number pattern.

   a   91, 94, 97, 100, …   b   201, 205, 209, 213, …

   c   60, 53, 46, 39, …   d   400, 200, 100, …

2   For each table, work out the pattern rule and the missing numbers.

Term number	1	2	3	4	?	10
Number of sides	4	5	6	7	8	?

Term number	1	2	3	4	5	?
Number of counters	0	1	2	3	?	79

Problem-solving

3   a   Mr Jones needs 3 456 blocks to build a wall.
        He already has 2 109 blocks. How many more does
        he need?

   b   Josh has 120 stickers. This is five times more than his friend James has.
        How many does James have?

   c   Jeremy has scored 312 runs this cricket season. Nick has scored 49 runs
        fewer than Jeremy. How many runs have they scored altogether?

Teaching notes

Perimeter

* Perimeter is a measure of the total length of the boundaries (sides) of a shape.
* You can measure perimeter, but you can also calculate perimeter if you are given the lengths of the sides.
* To calculate perimeter, you need to find the sum of the lengths of all sides of the shape.
* In regular shapes, the sides are the same length, so you can also calculate perimeter by multiplication. For example, a square with sides of 3 cm has a perimeter of 3×4 cm = 12 cm.

Area

* The area of a 2-D shape is the amount of space it covers.
* Area is measured in square units. At this level, the students need to count square units to determine the area. They should always give the units when they state the area, so the area is 9 squares, or 9 blocks, not just 9.
* Squared grid paper is very useful for exploring area: it allows students to see that very different shapes can have the same area.
* Remember that you can combine half squares to get a whole square.
* When you estimate area, you usually count the squares in which more than half is covered and ignore the squares in which less than half is covered. If squares are exactly half covered, you can combine them to get whole squares, or you can get an answer that is a mixed number (for example, an area of $9\frac{1}{2}$ squares).

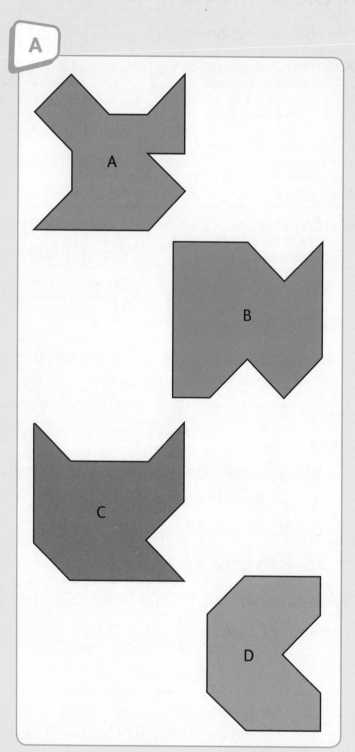

Rachael drew these shapes for an art project and outlined them in black. Which shape do you think has the longest outline? How could you find the length of the outline of each shape?

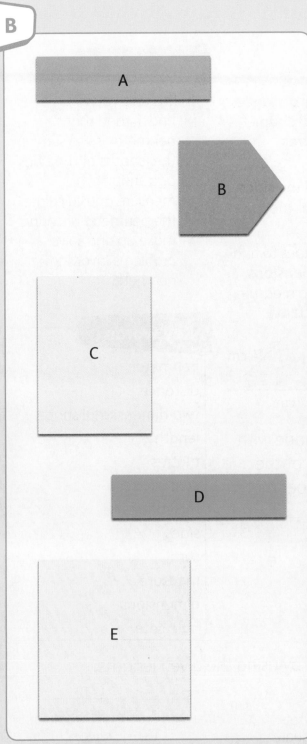

Look at these sticky notes. Which sticky note will take up the most space on a page? Romeo says that notes A and D will cover the same amount of space. How could you check whether he is correct?

Think, talk and write

A **Perimeter** *(pages 96–97)*

1 What units would you use to measure the distance around each of these objects?

  a your mathematics book

  b a cricket pitch

  c a stamp

2 Draw rough sketches of three different quadrilaterals that all have a perimeter of 16 cm.

B **Area** *(pages 98–100)*

Ms Daniels gives her students stickers when they do good work.

1 Which sticker will cover the most space on a page?

2 Which sticker will cover the least space on a page?

3 Which stickers will cover the same space on a page?

4 Which can you fit more of on a page: 'STAR!' stickers or 'BEST YET!' stickers?

A Perimeter

You already know that the **perimeter** is the total **distance** around the outside of a closed **two-dimensional shape**. Perimeter is given in units of **length** such as **metres**, **centimetres** and **millimetres**.

To calculate perimeter you add the lengths of all the **sides** of the shape together. So perimeter is the **sum** of the lengths of the sides of a shape.

You can **measure** the lengths of the sides of shapes to find the perimeter. If you are given a sketch with **dimensions** (measurements) written on it, you do not have to measure. You use the measurements you are given to **calculate** the perimeter.

The purple line is the perimeter of this triangle.

4 cm   4 cm

Perimeter = 4 cm + 4 cm + 7 cm = 15 cm   7 cm

Maths ideas

In this unit you will:
* understand that perimeter is the sum of the lengths of the sides of a shape
* measure and calculate the perimeter of shapes
* make up and solve problems involving perimeter.

Key words

perimeter

distance

two-dimensional shape

length

metres

centimetres

millimetres

sides

sum

measure

dimensions

calculate

1 Trace around the perimeter of each shape with your finger. Then measure the lengths of the sides of each shape and calculate the perimeter.

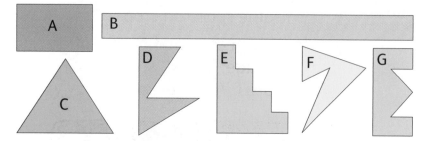

2 Calculate the perimeter of each shape by using the given lengths.

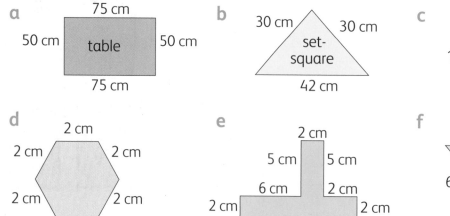

a   75 cm
50 cm   table   50 cm
75 cm

b   30 cm / 30 cm
set-square
42 cm

d   2 cm
2 cm / 2 cm
2 cm / 2 cm
2 cm

e   2 cm
5 cm / 5 cm
6 cm / 2 cm
2 cm / 2 cm
10 cm

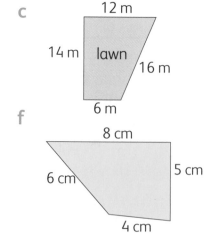

c   12 m
14 m   lawn   16 m
6 m

f   8 cm
6 cm   5 cm
4 cm

3 Calculate the perimeter of each of these objects.

a

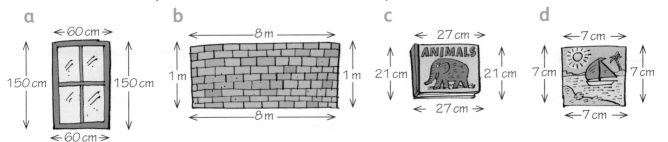

<60 cm>

150 cm 150 cm

<60 cm>

b

<———— 8 m ————>

1 m 1 m

<———— 8 m ————>

c

< 27 cm >

ANIMALS

21 cm 21 cm

< 27 cm >

d

<7 cm>

7 cm 7 cm

<7 cm>

4 What is the perimeter of each shape?

a Explain to your partner how you worked
out the missing dimensions.

b How can you use multiplication to find
the perimeter of rectangles and squares?
Share your ideas with your partner.

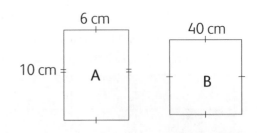

6 cm

10 cm A

40 cm

B

Problem-solving

5 Denison and Mike are making a square fishpond.
The perimeter of the pond is 8 metres.

a How long is each side of the pond?

b The boys want to put a line of tiles around the pond. The tiles they have
are squares with sides of 10 cm. How many tiles will they need?

c Draw a diagram to show your answers.

6 Merlene runs around her local park. The park is rectangular. One side is 93 m
long, the other is 14 m longer. Draw a labelled sketch of the park. Then use your
sketch to work out how far she will run if she goes round the park five times.

7 A centipede is 60 mm long from head to
tail. It moves along a rectangular path
from X to Y. How far will its head have
travelled when its tail is at point Y?

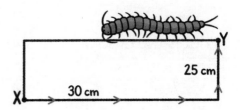

Y

25 cm

X 30 cm

What did you learn?

1 Make up two mathematics problems about perimeter and give them to your
partner to solve.

2 Draw a figure with a perimeter of 32 cm.

B Area

Maths ideas

In this unit you will:
* learn about area and what it means in mathematics
* count squares to work out the area of different shapes.

Key words

area   squares

space   square units

surface

Explain

Area is the amount of **space** or **surface** that a 2-D shape covers.

You can find the area of a shape drawn on a grid of equal **squares** by counting the number of squares the shape covers.

Area is measured in **square units**.
* Shape A has an area of 1 square unit.
* Shape B has an area of 2 square units.
* Shapes C and D have the same area – 3 square units.

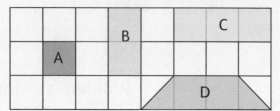

Sometimes there is not an exact number of whole squares inside the shape. You can then follow these steps to estimate its area:
* Count all the whole squares.
* Combine pairs of half squares to make whole squares.
* Count any parts that are bigger than half a square as a whole square.
* Ignore any parts that are smaller than half a square.
* Add up the squares you counted to get the total estimate of the area.

1   Write down the name of each shape and its area.

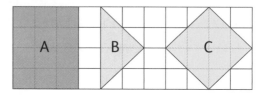

2   What is the area of each shape shown here?

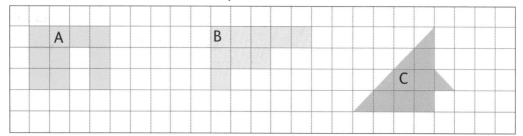

3   Clive and Troy are lost. They spell out this word on the ground using square tiles.

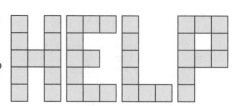

   a   What is the area of each letter in square units?
   b   How many square tiles did they use in all?

4 Shandra says these shapes all have the same area. Is she correct? How did you check?

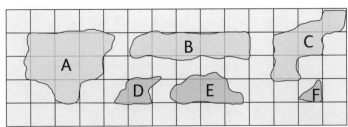

5 This map is drawn on a square grid. The area of island A is about 7 squares.

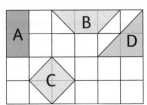

a Estimate the area of each of the other islands.

b Will a town with an area of 3 squares fit on island C?

c Estimate the area of the map that is covered by water.

Investigate

6 Your teacher will give you square grid paper and some small leaves.

a Use the grid to estimate the area of each leaf.

b Stick the leaves onto the grid and write the area next to each one.

Problem-solving

7 Curtis put some leaves on a square grid to estimate their area, but he couldn't see the squares through the leaves. Try to estimate the area of each leaf.

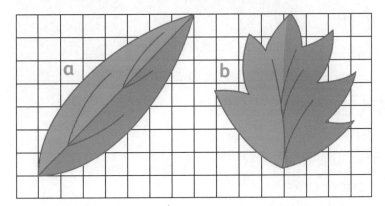

8   Some of the squares on these shapes were rubbed out by accident.

A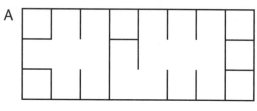

B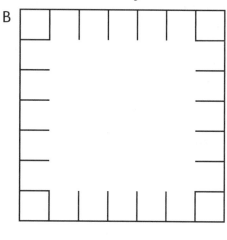

C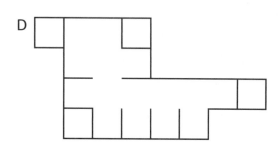

D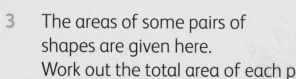

E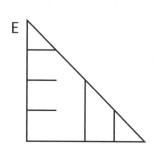

   a   Work out the area of each shape.

   b   Tell your partner how you worked out how many squares were missing.

9   Misha has a square with an area of 64 square units. He cuts a smaller square with an area of 25 square units out of the middle. What is the area of the shape that is left?

What did you learn?

1   What is the area of shape A?

2   Estimate the area of shape B.

3   The areas of some pairs of shapes are given here.
    Work out the total area of each pair.

   a   32 square units + 64 square units

   b   33 square units + 44 square units

   c   70 square units + 27 square units

   d   31 square units + 27 square units

A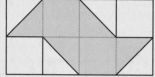

B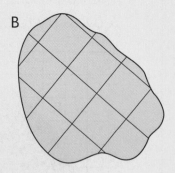

Topic 11 Review

Key ideas and concepts

Fill in the missing words in the sentences to summarise what you learnt in this topic.

1   The ____ is the distance around the outside of a figure.

2   To calculate the perimeter you use the length of each ____ and then ____ the measurements.

3   The area of a figure is the number of ____ that you need to ____ the surface.

4   You can find the area by ____ the number of squares in the shape.

Think, talk, write …

Mosaic patterns are often made from small square tiles. They can be used to make patterns on walls and floors. The picture shows part of a much larger mosaic pattern from the floor of a palace near Jericho. It is over 1300 years old.

How can you find the area of each colour in the pattern? Work with a partner.

The square tiles are 3 cm long. How can this help you find the perimeter of the pattern shown here?

Quick check

1   The word AREA has been formed using small squares.

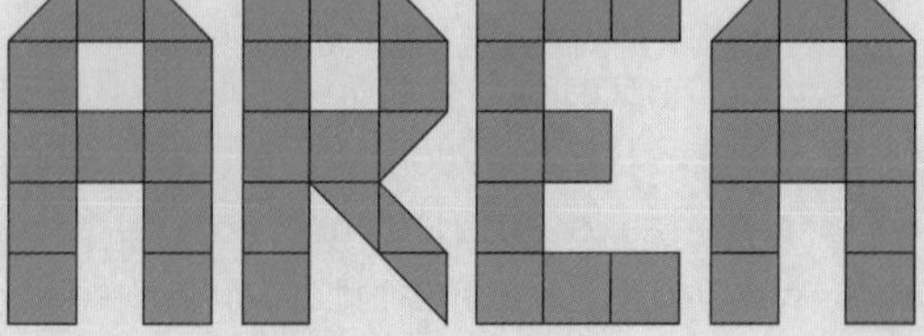

   a   What is the area of each letter in this word?

   b   Why did you only need to find the area of three of the letters?

   c   If each square has a length of 5 mm, what is the perimeter of the letter E?

Problem-solving

2   A quadrilateral has a perimeter of 40 cm. Three of its sides measure 6 cm, 6 cm and 14 cm.

   a   What is the length of the other side?

   b   What type of quadrilateral could this be?

A

Many of the objects you see around you are 3-D shapes. Look at this children's play structure. What solid shapes can you find? What other shapes can you see in the photograph?

Teaching notes

3-D shapes and their parts

* Solid objects are three-dimensional (3-D) shapes, because they have three dimensions that you can measure: length, breadth and height.
* Flat surfaces on 3-D shapes are called faces.
* Some 3-D shapes don't have faces but have curved surfaces instead. Ball shapes, for example, have only one curved surface, and no faces.
* Where two faces meet they form an edge.
* Any corner that is formed where three (or more) faces of a 3-D shape meet is called a vertex (plural: vertices).
* The point of a cone is also called its vertex.

Naming 3-D shapes

* Solids are named according to the number and shape of their faces.
* A cube has six square faces.
* A cuboid or rectangular prism has six rectangular faces (some may be square).
* A cone has a flat circular base and a curved surface that forms a point (the vertex).
* A cylinder has two flat circular end faces and a curved surface.
* A sphere (or ball shape) has one curved surface.

B

Look at these differently shaped dice. Which have more faces than a normal six-sided dice? Which dice have flat faces? Which have curved faces? Choose one type and describe it to your partner. Let them try to guess which one it is.

Think, talk and write

A **Naming 3-D shapes** *(page 104)*

1 Can you name these 3-D shapes?

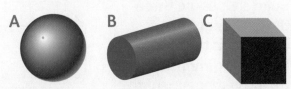

2 Tell your partner where you would find two examples of real objects that are like each shape.

B **Properties of 3-D shapes** *(pages 105–106)*

1 Look at the shapes above again.

   a Which shapes can roll?

   b Which shapes can stack?

   c Which shapes can roll and stack?

2 Do you know the mathematical name for these parts of shapes?

   a   b

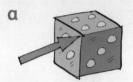   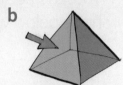

A Naming 3-D shapes

Explain

Solid shapes are **three-dimensional (3-D)**.
This means the shape has three **dimensions** (measurements): length, width and height.

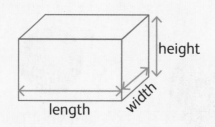

3-D is short for three-dimensional.

Do you know the names of these 3-D shapes?

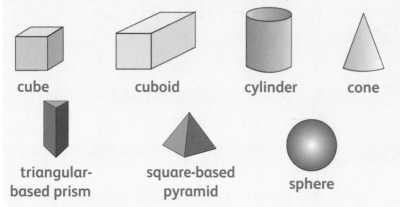

cube   cuboid   cylinder   cone

triangular-based prism   square-based pyramid   sphere

Key words

solid	cylinder
three-	cone
dimensional	sphere
(3-D)	triangular-
dimensions	based prism
cube	square-based
cuboid	pyramid

1   Write the names of the seven types of 3-D shapes in a list. Match each 3-D shape below with its name. Write the letter of each shape next to its name.

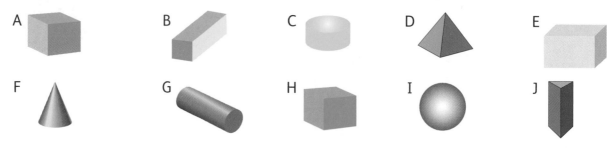

2   How can you tell the difference between:

   a   cubes and cuboids?   **b**   cones and cylinders?   **c**   prisms and pyramids?

3   Work in pairs. Choose one 3-D shape. Find as many examples as you can of real objects that are this shape. Share your list with the class.

What did you learn?

1   What shape is the outside of this blue block?

2   What shape has been cut from the centre of the block?

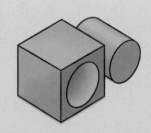

B Properties of 3-D shapes

Explain

The parts of 3-D shapes are given special mathematical names.

A **face** is a **flat** side of a solid.

An **edge** is where two faces of a solid meet.

A **vertex** is a point where three or more faces of a solid meet. The plural of vertex is **vertices**.

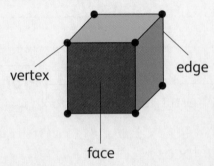

vertex   edge   face

This cube has 6 square faces, 8 vertices and 12 edges.

A cylinder has two **circular** faces (the flat circles) and a **curved** surface. Since the two faces do not meet, a cylinder has no edges and no vertices.

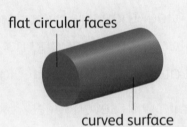

flat circular faces

curved surface

Maths ideas

In this unit you will:
* revise the mathematical names of parts of shapes
* learn about the special properties of different 3-D shapes.

Key words

face

flat

edge

vertex

vertices

circular

curved

1   Look at these shapes.

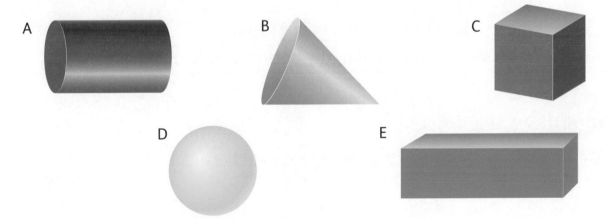

A
B
C
D
E

Name the shapes that have:

a   no flat faces   **b**   only one vertex   **c**   two circular faces

d   six rectangular faces   **e**   no edges.

2 Read the information in the table carefully. Write the letters A to E in your book and name the shape that matches each set of properties.

Shape	Total number of surfaces	Number of flat faces	Number of curved surfaces	Number of edges	Number of vertices
A	6	6	0	12	8
B	6	6	0	12	8
C	3	2	1	2	0
D	2	1	1	0	1
E	1	0	1	0	0

Problem-solving

3 Draw and name each of these solids from their descriptions.

   a   a box shape with only rectangular faces

   b   a solid with two square faces and four rectangular faces

   c   a tube shape that is wider than it is high

   d   a shape with no flat faces, no edges and no vertices

   e   a shape with a point at the top and a circular base

4 Is it possible to draw a solid with seven faces? Explain your answer.

5 Answer these questions about 3-D shapes.

   a   Can you use a cube to trace a square?

   b   Can you draw a circle using a cone?

   c   Which shape has faces that are all identical?

   d   Which shapes can you draw by tracing around a cylinder?

What did you learn?

Draw a shape like this one in your book.

1 Write the name of the shape.

2 Label it to show an edge, a face and a vertex.

3 How many edges does it have in all?

4 What is the sum of the faces and the vertices?

5 Name four everyday objects that are this shape.

Topic 12 Review

Key ideas and concepts

Make a poster to summarise what you learnt about 3-D shapes in this topic.

1   Cut out pictures of everyday objects shaped like cubes, cuboids, cones, cylinders and spheres. (Or draw your own pictures.)

2   Stick the pictures onto a poster and label them to show the five types of shapes and their properties.

Think, talk, write …

Nadia built this model using 18 wooden cubes.

How many cubes will show:

1   3 faces?

2   2 faces?

3   1 face?

4   4 faces?

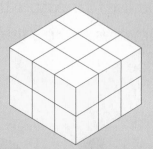

Quick check

1   Write the mathematical name for a 3-D shape that is shaped like a:

    a   ball   **b**   shoe box   **c**   dice   **d**   can of beans.

2   Look at these shapes.

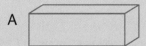 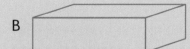

    a   What is the correct name of each shape?

    b   Which shape has six rectangular faces?

    c   Where would you find a vertex on these shapes?

Test yourself (2)

Explain

Complete this test to check that you have understood and can manage the work covered in Topics 1 to 12.

Revise any sections that you find difficult.

1 Write each number.

    a four thousand three hundred and nine

    b eight thousand nine hundred and twenty

    c six thousand and six

2 Which of these units could you use to measure the length of a classroom?

    a grams   **b** centimetres   **c** litres   **d** metres

3 **a** What fraction of each shape is shaded?

    b Write the fraction of each shape that is unshaded. Then give an equivalent fraction for each.

4 Estimate and then measure the length of each line segment in millimetres.

5 How many grams are in:

    a 1 kg?   **b** 1 kg 800 g?   **c** 4.5 kg?

6 Nadia had two bags. One had a mass of 15 kg, the other had a mass of 9 kg. What was their total mass?

7 **a** Match each shape to its name.

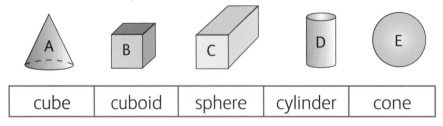

| cube | cuboid | sphere | cylinder | cone |

    b How are cubes and cuboids similar?
        Write two things that are the same.

8 Determine any missing measurements and then calculate the perimeter of each shape.

a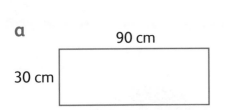

b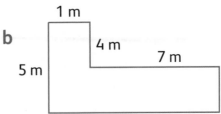

9 If the number pattern 7, 10, 13, 16, … were continued, which of these numbers would be terms in the pattern?

a   186   b   187   c   188   d   189

10 Nicola has 63 mangoes. She shares them equally into four bags. How many are left over?

11 One computer cost $2 350. How much would it cost if a school bought 3 at that price?

12 Find four numbers that will add up to 600. Two of the numbers must be two-digit numbers. One must be a three-digit number. The fourth number can have any number of digits.

13 An albatross has a wingspan of 315 cm. A seagull has a wingspan of 107 cm. What is the difference in wingspan between the two birds?

14 Find the next three numbers in each pattern. Then write the rule for the pattern.

a   723, 623, 523, ___, ___, ___   b   2, 4, 8, 16, ___, ___, ___

15 What is the missing value in each number sentence?

a $2007 - 200 = \square$ b $\square + 132 = 341$

c $\square \times 56 = 224$ d $\square \div 2 = 125$

16 Rewrite each set of fractions in order from the smallest to the greatest.

a $\dfrac{1}{2}$ $\dfrac{1}{4}$ $\dfrac{1}{8}$ $\dfrac{1}{7}$ $\dfrac{1}{3}$

b $\dfrac{1}{3}$ $\dfrac{1}{6}$ $\dfrac{1}{8}$ $\dfrac{2}{3}$ $\dfrac{2}{8}$

17 Find the area of each shape in square units.

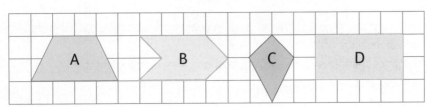

A

Mrs Joyson buys a bag of oranges and squeezes them to make fresh juice for her breakfast. She uses $\frac{3}{9}$ of the bag on Monday and $\frac{2}{9}$ of the bag on Tuesday. What fraction of the bag did she use on the two days?

Teaching notes

Addition with fractions

* When fractions have the same denominators, you can add them by thinking of them as 'pieces'. For example, to you add $\frac{2}{5}$ and another $\frac{2}{5}$, ask yourself how many fifths in all. The answer is $\frac{4}{5}$. The denominator does not change. You are adding two lots of fifths, so the answer is in fifths.

* To add fraction with different denominators, use equivalent fractions to make the denominators the same. For example, to add $\frac{1}{4}$ and $\frac{1}{2}$, write $\frac{1}{2}$ as $\frac{2}{4}$ and then add the numerators: $\frac{1}{4} + \frac{2}{4} = \frac{3}{4}$.

* When you add fractions you may get a result that is an improper fraction or a mixed number. $\frac{3}{4} + \frac{3}{4} = \frac{6}{4}$. The answer $\frac{6}{4}$ is the same as 1 whole and $\frac{2}{4}$ or $1\frac{1}{2}$.

Subtraction with fractions

* As with addition, you use fractions with the same denominators when you subtract, and you can think of them as 'pieces'. $\frac{5}{6} - \frac{1}{6}$ means five sixths less one sixth, which is $\frac{4}{6}$.

* To subtract fractions with different denominators, use equivalent fractions. $\frac{5}{6} - \frac{1}{3}$ is equivalent to $\frac{5}{6} - \frac{2}{6} = \frac{3}{6}$.

* Answers such as $\frac{4}{6}$ and $\frac{3}{6}$ can be simplified to get $\frac{2}{3}$ and $\frac{1}{2}$.

1234

A Adding fractions (pages 112–113)

1 This cake has been cut into 8 equal pieces.

Jonelle eats a piece and her cousins eat four pieces.

   a What fraction of the cake has been eaten?

   b How could you write this as a sum?

2 Find three fractions in this group that are equivalent.

$\frac{1}{3}$	$\frac{1}{2}$	$\frac{3}{4}$	$\frac{4}{8}$	$\frac{2}{5}$	$\frac{5}{10}$

B Subtracting fractions (pages 114–115)

1 Look at these folded sheets of paper.

2 What fraction of each sheet will be left if you cut off two pieces?

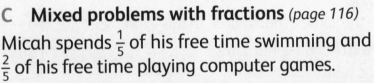

C Mixed problems with fractions (page 116)

Micah spends $\frac{1}{5}$ of his free time swimming and $\frac{2}{5}$ of his free time playing computer games.

1 What fraction of his free time does he spend doing these two things?

2 What fraction of his free time does he spend doing other things?

B

Glen cuts a slice of orange into equal parts like this. What fraction of the slice is each part? He eats 4 of the parts. What fraction of the slice will he have left?

A Adding fractions

Explain

You can add **fractions** much like any other numbers.

Look at this example to help you remember how to **add** fractions when the **denominator** is the same.

$$\frac{3}{8} + \frac{2}{8} = \frac{5}{8}$$

| $\frac{1}{8}$ | $\frac{1}{8}$ | $\frac{1}{8}$ | | | | | | + | $\frac{1}{8}$ | $\frac{1}{8}$ | | | | | |

You are adding eighths to eighths, so your answer is also in eighths. To find the **total** number of eighths, simply add the **numerators**.

You can also add fractions to **whole numbers**. Your answer will be a **mixed number**.

$$3 + \frac{2}{5} = 3\frac{2}{5}$$

When you do **addition** of fractions with different denominators, you can use **equivalent** fractions to make the denominators the same.

$\frac{2}{3} + \frac{1}{6}$ Change one fraction so that both fractions have the same denominator. In this example you will change $\frac{2}{3}$ into sixths:

$$\frac{2}{3} \times \frac{2}{2} = \frac{4}{6}$$

So the number sentence is: $\frac{4}{6} + \frac{1}{6} = \frac{5}{6}$

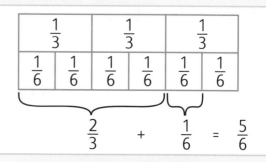

1   Add.

a $\frac{3}{5} + \frac{1}{5}$ b $\frac{2}{10} + \frac{5}{10}$ c $\frac{1}{6} + \frac{3}{6}$

d $\frac{1}{7} + \frac{5}{7}$ e $\frac{2}{8} + \frac{2}{8}$ f $\frac{1}{7} + \frac{4}{7}$

g $\frac{4}{9} + \frac{3}{9}$ h $\frac{1}{12} + \frac{6}{12}$ i $\frac{1}{5} + \frac{1}{5}$

j $\frac{4}{11} + \frac{3}{11}$ k $\frac{8}{9} + \frac{1}{9}$ l $\frac{1}{12} + \frac{9}{12}$

2 Work out the **sum** of:

a $2 + \frac{1}{4}$ b $3 + \frac{2}{9}$ c $10 + 3 + \frac{1}{4}$ d $12 + 3 + \frac{4}{9}$

e $4 + 3 + \frac{1}{3} + \frac{1}{3}$ f $2 + 5 + \frac{2}{9} + \frac{7}{9}$ g $3 + \frac{1}{4} + 5 + \frac{1}{4}$

3 Add. Show your working. Use a fraction wall if you need to.

a $\frac{1}{3} + \frac{1}{6}$ b $\frac{2}{5} + \frac{3}{10}$ c $\frac{1}{4} + \frac{1}{8}$ d $\frac{2}{4} + \frac{5}{8}$

e $\frac{1}{2} + \frac{1}{6}$ f $\frac{3}{4} + \frac{1}{12}$ g $\frac{5}{8} + \frac{1}{4}$ h $\frac{2}{5} + \frac{3}{10}$

i $\frac{2}{5} + \frac{2}{15}$ j $\frac{5}{9} + \frac{1}{3}$ k $\frac{1}{2} + \frac{2}{10}$ l $\frac{2}{3} + \frac{1}{12}$

Problem-solving

4 Rihanna spent $\frac{2}{9}$ of her money at the beach and $\frac{1}{3}$ of her money at the movies.

   a What fraction of her money did she spend?

   b What fraction of her money did she have left?

5 Three bags of mangoes have masses of $\frac{1}{4}$ kg, $\frac{1}{2}$ kg and $\frac{1}{8}$ kg. What is their combined mass?

Challenge

6 In this table, the sum of the fractions in each row and each column is the same.

   a Work out the value of A, B, C and D.

   b Write each fraction from A to D in its simplest form.

A	$\frac{1}{4}$	$\frac{2}{3}$
B	$\frac{1}{2}$	$\frac{7}{12}$
C	$\frac{3}{4}$	D

What did you learn?

1 A cake is cut into 10 equal parts. Micah eats 3 parts, Jess eats 1 part and Nicky eats 2 parts. What fraction of the cake have they eaten?

2 Sarah has 4 packets of elastic bands. Each packet is $\frac{1}{8}$ full. Does she have more or less than half a packet?

B Subtracting fractions

Explain

You can **subtract fractions** much like any other numbers when the **denominators** are the same.

$$\frac{5}{6} - \frac{2}{6} = \frac{3}{6}$$

You are subtracting sixths from sixths, so your answer will also be in sixths.

Last year you learnt how to subtract a fraction from a **whole**. This example will remind you how to do this:

$1 - \frac{1}{7}$

Remember 1 is **equivalent** to $\frac{7}{7}$.
So you can simply subtract the **numerators**.

$$\frac{7}{7} - \frac{1}{7} = \frac{6}{7}$$

When the denominators are different, you can use equivalent fractions to subtract.

The example shows you how to do this:

$\frac{5}{9} - \frac{1}{3}$

Change the thirds to make an equivalent fraction with a denominator of 9:

$$\frac{1}{3} \times \frac{3}{3} = \frac{3}{9}$$

So the **subtraction** number sentence is:

$\frac{5}{9} - \frac{3}{9} = \frac{2}{9}$ You subtract the numerators to find the **difference**.

$\frac{5}{9} - \frac{1}{3} = \frac{2}{9}$

Maths ideas

In this unit you will:
* use diagrams to understand how to subtract fractions
* subtract fractions using diagrams and the fraction wall
* write a number sentence for subtracting fractions and find the difference
* solve problems involving subtraction of fractions.

Key words

subtract	equivalent
fractions	numerators
denominators	subtraction
whole	difference

1   Subtract.

a $\frac{8}{9} - \frac{3}{9}$

b $\frac{5}{6} - \frac{2}{6}$

c $\frac{9}{11} - \frac{3}{11}$

d $\frac{7}{8} - \frac{3}{8}$

e $\frac{15}{20} - \frac{11}{20}$

f $\frac{5}{8} - \frac{4}{8}$

g $\frac{17}{20} - \frac{12}{20}$

h $\frac{9}{10} - \frac{5}{10}$

i $\frac{3}{5} - \frac{3}{5}$

j $\frac{14}{20} - \frac{5}{20}$

2   The cards show what fraction of an hour some students spent doing
    their homework.

Maya	Josh	Maria	Peter	Zara
$\frac{2}{5}$	$\frac{5}{8}$	$\frac{3}{7}$	$\frac{7}{10}$	$\frac{2}{9}$

   a   What fraction of an hour does each student have left after they've done
       their homework?

   b   Which students have more than half an hour left?

3   Find the difference between each pair of fractions.

   a $\frac{3}{4} - \frac{5}{8}$ b $\frac{2}{3} - \frac{2}{6}$ c $\frac{3}{4} - \frac{1}{2}$ d $\frac{1}{3} - \frac{1}{6}$ e $\frac{2}{3} - \frac{3}{6}$

   f $\frac{3}{4} - \frac{1}{8}$ g $\frac{7}{8} - \frac{1}{2}$ h $\frac{5}{9} - \frac{1}{3}$ i $\frac{10}{12} - \frac{1}{4}$ j $\frac{9}{10} - \frac{1}{2}$

   k $\frac{8}{9} - \frac{2}{3}$ l $4 - \frac{2}{3}$ m $\frac{7}{9} = \frac{2}{3}$ n $\frac{7}{8} - \frac{1}{4}$ o $\frac{2}{5} - \frac{3}{10}$

Problem-solving

4   Jayden has $\frac{7}{8}$ of his pocket money left. On Saturday he
    spends half of it. What fraction of his money does he have left?

5   Mrs Nixon makes curry to sell at a church fete. She mixes $\frac{1}{2}$ kilogram of
    chicken with $\frac{1}{4}$ kilogram of vegetables. Later she cooks $\frac{1}{3}$ kilogram of rice.
    What is the total mass of the food?

6   Jonas has $\frac{2}{5}$ m of rope and Josh has $\frac{4}{9}$ m of rope.
   a   What is the longest length they can make if they lay the ropes end
       to end?
   b   Josh's mother cuts $\frac{1}{4}$ metre from his piece of rope. What is he left with?

What did you learn?

1   Explain why you make fractions with the same denominators when you have
    to add or subtract fractions.

2   What would you need to do to:

   a   add $\frac{1}{2}$ and $\frac{1}{3}$?   b   subtract $\frac{1}{3}$ from $\frac{1}{2}$?

C Mixed problems with fractions

You can use a bar model to see the information you already have and what it is you need to work out when you have to solve problems involving **fractions**.

Example

Sondra has $\frac{2}{3}$ of a pizza and Nisha has $\frac{4}{6}$ of a pizza. How much pizza do they have altogether?

Draw two bars to represent the two pizzas.

Divide the first bar into thirds and the second into sixths and show each girl's pizza.

Now you can see that you need to add.

$$\frac{2}{3} + \frac{4}{6} = \frac{4}{6} + \frac{4}{6} = \frac{8}{6},$$

which is $1\frac{2}{6}$ or $1\frac{1}{3}$ pizzas.

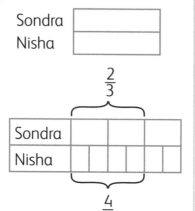

Maths ideas

In this unit you will:
* use what you have learnt in this topic to solve problems involving fractions
* apply suitable strategies to help you solve fraction problems.

Key words

fractions

left over

Shaun, James and Nick each had the same sized pizza. Shaun ate $\frac{7}{12}$ of his, James ate $\frac{3}{8}$ of his and Nick ate $\frac{3}{4}$ of his.

1   Who ate the most pizza?

2   How much pizza was **left over** in total?

What did you learn?

Zara spent $\frac{1}{5}$ of her money on a book. If the book cost $15. How much money did she have to start with?

Topic 13 Review

Key ideas and concepts

Complete these sentences to summarise the skills you learnt in this topic.

1 To add fractions with different denominators I would _____.

2 To subtract a fraction from another fraction I would _____.

3 When a problem involves fractions I would _____.

Think, talk, write ...

1 Look at this bar model.

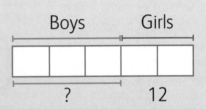

   a Write a problem it could represent.

   b How would you solve it?

2 What did you find the easiest in this topic? What did you find the most challenging? Write about this in your maths journal.

Quick check

1 Try to do all these calculations mentally.

   a $\dfrac{3}{5} + \dfrac{1}{5}$
   b $\dfrac{9}{20} + \dfrac{3}{20}$
   c $\dfrac{1}{4} + \dfrac{3}{4}$
   d $5 - \dfrac{3}{4}$

2 Calculate.

   a $\dfrac{1}{6} + \dfrac{1}{12}$
   b $\dfrac{1}{5} + \dfrac{7}{10}$
   c $\dfrac{1}{4} + \dfrac{2}{5}$
   d $\dfrac{3}{4} - \dfrac{1}{2}$

   e $\dfrac{3}{5} - \dfrac{1}{10}$
   f $\dfrac{4}{10} - \dfrac{1}{5}$
   g $\dfrac{4}{5} + \dfrac{2}{15}$
   h $\dfrac{1}{3} + \dfrac{1}{12}$

3 An empty bottle has a mass of $\dfrac{1}{10}$ of a kilogram. When it is filled with water, it has a mass of $\dfrac{11}{12}$ of a kilogram. What is the mass of the water?

4 Nisha pours $\dfrac{3}{7}$ of a litre of water from a two litre container. How much water will be left in the container?

5 $\dfrac{1}{12}$ of a circle is yellow. $\dfrac{1}{3}$ of it is green. The rest is blue. What fraction of the circle is blue?

Teaching notes

Time

* Students should be able to tell time in five-minute intervals. They will extend this to single minutes this year.

* Students need to be familiar with the different conventions for writing time: three o'clock in the afternoon can be written as 3 o'clock, 3 p.m., 3:00 p.m. and 15:00. From midnight to noon we use a.m., from noon to midnight we use p.m. For noon and midnight, we usually do not add a.m. or p.m.

* Duration indicates how much time an event takes. You can work this out by counting on (or back) in hours and minutes. Time between events (say, between the start and end of a race) can be worked out in the same way.

* Units of time are not decimal, but they do have equivalents. 60 seconds equal one minute, and 60 minutes make up one hour. Seven days make one week, twelve months make one year. Students should be able to do simple conversions between these units.

Money

* Mixed amounts in dollars and cents are written using a decimal point, for example, $5.65. Students will see these in daily life, even though they do not work with decimal notation yet.

* Students need to be able to calculate money amounts and have to learn the correct vocabulary for money problems.

A

These two watches show time in different ways. The watch on the left uses hands to show the time on the face. How does the watch on the right show the time? Where else do you find time shown in these two ways? Which method of showing time do you find easier to read and understand? Why?

B

BIG SUMMER SALE!

ALL ITEMS REDUCED!!!

½ PRICE OR LESS!

Read the sign. What information does it give you? What do the words 'sale' and 'reduced' mean? What does it mean if something is $\frac{1}{2}$ price or less?

Think, talk and write

A Work with time *(pages 120–122)*

1 How long is your school day? What time does it start and end?

2 What time is midnight? How can you write this time?

3 Which is longer, 45 minutes or 1 hour?

4 How can you work out how long something takes?

5 How would you add 1 week and 3 days to 3 weeks and 6 days?

B Work with money

(pages 123–126)

Work in groups and talk about the coins and notes that are used in Trinidad and Tobago.

1 Make a list of all the coins used. Next to each one, write the colour of metal it is made from. What do you notice?

2 Make a list of the values and colours of the notes in use.

3 What is the note with the lowest value? How can you make that amount in coins?

4 What is the note with the highest value? Can you think of something that costs about that much money?

A Work with time

Explain

Hours, **minutes** and **seconds** are units of time.

1 minute = 60 seconds   1 hour = 60 minutes

$\frac{1}{2}$ hour = 30 minutes $\frac{1}{4}$ hour = 15 minutes

Watches and clocks have an **hour hand** (the short hand) and a **minute hand** (the longer hand).

You can work out the number of minutes past the hour on a clock by starting at 12 and counting on in fives until you get close to the number the minute hand is pointing to. Then you count in ones to find the exact minute.

This clock shows 17 minutes past one.

Now it shows 8 minutes to two.

A clock with 12 divisions that uses hands to show the time is called an **analogue** clock.

A clock that only uses numbers to tell the time is called a **digital** clock.

Most electronic clocks are digital, including clocks on mobile phones and computers.

This digital clock shows that it is 14 minutes past 9.

You write **a.m.** for times that are after midnight and before noon (midday).

You write **p.m.** for times that are after noon (midday) and before midnight.

9:14 a.m. means 14 minutes past 9 in the morning.

9:14 p.m. means 14 minutes past 9 at night.

Maths ideas

In this unit you will:
* tell and write time to the minute in different ways, using analogue and digital clocks
* write dates in different ways
* learn how different units of time are related to each other
* solve problems to work out how long things take
* understand and use time vocabulary.

Key words

hours

minutes

seconds

hour hand

minute hand

analogue

digital

a.m.

p.m.

days

weeks

months

date

year

1   Work with a partner. Say each time aloud. Then write what it would be if it was shown on a digital clock.

a    b    c    d    e

f    g    h    i    j

2   Write each time correctly using a.m. or p.m.

a   8 in the morning
b   8 at night
c   3 in the morning
d   3 in the afternoon
e   six thirty in the morning
f   nine before 3 in the afternoon.

3   Complete the following.

a   2 hours = ☐ minutes
b   6 hours = ☐ minutes
c   12 hours = ☐ minutes
d   180 minutes = ☐ hours
e   15 minutes = ☐ hours
f   30 minutes = ☐ hours

4   These clocks show six times in the afternoon.

a   Write the times in order from the earliest to the latest.
b   Which time is closest to quarter to 4?

Challenge

5   Elene walked on the beach for 1 hour and 23 minutes.
James walked on the beach for 83 minutes.
Who spent more time walking on the beach?

6   How many **days** are there in:
a   1 **week**?   b   four weeks?   c   the **month** of September?

7   Write these dates in number format.
a   tomorrow's date   b   the **date** of your birthday next **year**

8   How many days is it from:
a   9 October to 24 October?   b   14/02 to 28/02?

Explain

To work out how long something takes, you can count on from the time something starts to the time when it ends.
Try counting in 5s, 10s or 15s.

1   Work out how long it is from each start time to end time.
    a   3:00 a.m. → 7:45 a.m.   b   3:00 a.m. → 4:00 p.m
    c   6:16 a.m. → 8:16 a.m.   d   9:20 p.m. → 11:40 p.m.
    e   8:15 p.m. → 11:20 p.m.   f   10:05 a.m → 11:25 p.m.

2   How long did these games last? Work out the answers.
    a   The tennis game started at 2:00 p.m. and finished at 3:45 p.m.
    b   The baseball game started at 6 p.m. and finished at quarter past 8 the same night.

3   Your family wants to travel from Barbados to London. There is a flight that leaves Barbados at 11:00 p.m. The flight takes $12\frac{1}{2}$ hours. Your friend says that the flight will arrive in London at 4:30 p.m. the next day. Is that correct?

4   The movie at a cinema starts at 4 p.m. It takes Paula 20 minutes to bathe, 10 minutes to get dressed and 25 minutes to walk to the cinema. What is the latest time she can get into the bath if she wants to be on time for the movie?

Investigate

5   Use a dictionary, the internet or any other source of information and find out what each of these words means:

decade	century	millennium	leap year	anniversary

Tell your group what you found out.

What did you learn?

1   A class started at 12:30 p.m. and ended at 2:15 p.m. How long was the class?

2   Mr Butler makes a cake. It takes him 45 minutes. He starts at 11:15 a.m. What time does he finish?

B Work with money

Explain

Coins and notes have different values shown on them. The value of a coin, note or stamp is called its **denomination**. The denomination of a ten-dollar note is ten dollars.

You can combine **coins** and **notes** of different denominations to make money amounts. These three children have all made twenty dollars.

$10 + $5 + $5 = $20 $4 \times $5 = $20 $10 + $5 + (5 \times $1) = $20

Can you think of one other way of making up a total of $20?

Maths ideas

In this unit you will:
* describe and recognise the coins and notes used in your country
* read and write amounts of money
* make amounts of money using different combinations of coins and notes
* work with prices and calculate total amounts and change due
* discuss how to deposit money into a bank account and how to withdraw money from it.

1   How many ways can you find to make each amount using only $5, $10, $20 and $50 dollar notes? Write your combinations as calculations, like the ones above. You can use as many of each denomination as you want.

   a $25   b $45   c $500
   d $75   e $100   f $90

Key words

denomination
coins   prices
notes   cost
amount due   change
estimate   deposit
calculate   withdraw

2   Write two ways of making each amount using any combination of coins.

   a $2   b $1
   c $1.25   d $3.75

3   Choose one coin used in Trinidad and Tobago. Describe it to your partner without saying its denomination. Try to guess which coin your partner is describing.

You can add, subtract, multiply and divide amounts of money.

When you are working with amounts in dollars and cents, it is useful to round them off to the nearest whole dollar to **estimate** before you **calculate**.

To round off money amounts, look at the digit after the decimal point. If it is five or greater, round the dollars up to the next dollar. If it is less than five, leave the dollar amount unchanged and write zeros as place holders in the money amount.

Example

A book costs $3.66 and a pen costs $1.87. What is the **amount due** if you buy one of each?

Estimate:
$4.00 + $2.00 = $6.00

Real cost:
$3.66
$1.87
$5.53

To add or subtract amounts with cents, make sure you put the places below each other.

4   Look at the pictures and **prices** of all these items.

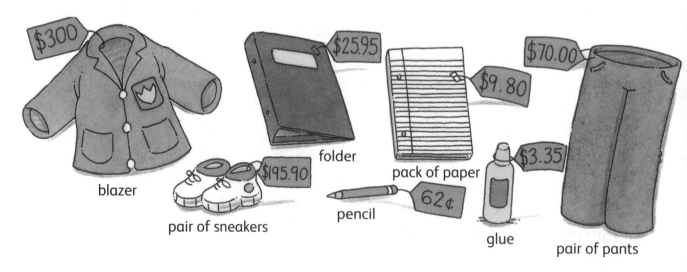

blazer — $300
folder — $25.95
pack of paper — $9.80
pair of sneakers — $195.90
pencil — 62¢
glue — $3.35
pair of pants — $70.00

Estimate and then calculate the total **cost** of each set of objects.

a   2 pencils, 1 pack of paper and 1 bottle of glue

b   1 folder, 1 pack of paper and 1 pencil

c   2 folders, 1 pack of paper and 3 bottles of glue

d   2 folders, 9 pencils, 2 packs of paper, 2 bottles of glue

e   2 pairs of pants, 1 pack of paper, 3 folders, 1 blazer

f   2 pairs of pants, 1 blazer and a pair of sneakers

5   The items in the picture below are on sale. The reduced sale price of each item is given.

fridge was
$1 064.00
reduced to
$1 024.00

living room
sofa was
$4 995.95
reduced to
$4 720.00

cooker was
$1 805
reduced to
$1 760.00

radio was
$164.00
reduced to
$156.00

a   Find the total cost of three refrigerators and a stove at the old price.

b   Find the total cost of a radio and a stove at the reduced price.

c   How much do you save by buying a radio and a stove at the reduced prices?

d   How much cheaper is the sofa at the reduced price?

e   How much would you save altogether if you bought a fridge and a sofa on sale?

Explain

When you buy things, you often hand over slightly more money than you need to pay because you don't have the exact amount.

For example, Mary wants to buy a book that costs $7.50. She pays with a $10 note and gets $2.50 back as **change**.

The person selling the items then works out how much extra you have given them and gives you back the balance as change.

In most stores, the cash register calculates this amount exactly.

Example

Subtract in parts. Use a number line if you need to.

$10 – $7.50 = $10 – $7 – 50¢ = $3 – 50¢ = $2.50

$$-50¢ \qquad -$7$$

$2.50 $3 $10
change

Add on in parts. Use a number line if you need to.

$7.50 + 50¢ + $2.00 = $10.00

$$+50¢ \qquad +$2 \longrightarrow $2.50 \text{ change}$$

$7.50 $8 $10

You can work out the change due to you by subtracting or adding on.

6   What change will you get if you buy this shirt and give the cashier:

   a $15?   b $20?   c $50?

7   What change will you get if you pay for this skirt by giving the salesperson:

   a $25.00?   b $30.00?   c $50.00?

Problem-solving

8   Susan bought a book that cost $8.70. She received $1.30 change. What amount did she give the cashier?

9   Peter bought a hat and received two 25¢ coins as change. He paid with a $20 note. What did the hat cost?

Investigate

10   Today you can put money into a bank account using an automatic teller machine and a bank card. You put the money into an envelope and deposit it into the machine. You can also take money out of your bank account using an automatic teller machine and your bank card.

In the past, people could not do this – they had to go into the bank to **deposit** or **withdraw** money.

Interview some older people in your family or community to find out how they used paper deposit and withdrawal slips to put money into the bank and to get money back out of the bank.

Tell the class what you found out.

What did you learn?

1   Mario has $4 in coins. What coins could he have?

2   Sue has three notes in her pocket. They add up to less than $30. What notes could she have?

3   An item cost $39.50 on sale. If the original price was $12.50 more, what was it?

4   a   What would it cost to buy 5 pencils that cost 90¢ each?
    b   How much change would you get if you paid for the pencils using a $10 note?

Topic 14 Review

Key ideas and concepts

Copy and complete this mind map to summarise what you learnt in this topic.

Write at least five points for each sub-topic.

Time

MEASUREMENT

Money

Think, talk, write ...

1   Work with a partner. Make up a test with two questions from each unit (time and money).

Exchange with another pair and complete each other's test.

Hand back the tests and mark them.

2   Why do you think that dollar notes come in denominations of 1 or multiples of 5 or 10? Why don't we have $6 notes or $9 notes?

Quick check

1   Write these times in order from the earliest to the latest.

1:30 p.m.   5:45 a.m.   3:20 p.m.   9:05 p.m.

2   What time is shown on each clock?

3   Mario and Kamaya are working on a project. They estimate it will take 5 lessons to complete it. If each lesson is 30 minutes long, how long will they spend on the project?

4   A store offers a $750 reduction on a fridge that cost $2 299. What would the reduced price be?

5   How much change will you get from $20 if you buy items costing $2.50, $3.75 and $10.30?

Teaching notes

Frequency tables

* The data in each category of a frequency table can be shown visually on a graph.
* The type of graphs and the scale shown on it depend on the data in the table.

Pictographs

* A pictograph is a graph that uses symbols or pictures to show data in different categories.
* Pictographs should have a heading and key to say what the symbols represent.

Bar graphs

* A bar graph is a graph that uses bars, with a space between them to show the number of data items in different categories. The bars should be the same width and the spaces between them should be equal.
* Bar graphs need a scale on one axis so that you can work out the number of data items (frequency) for each bar. The other axis has the categories for each bar.
* Bar graphs can be horizontal or vertical.
* Bar graphs do not need a key, but they do need a heading.

Making sense of graphs

* The heading, key and scale of a graph give important information about the graph.
* You can read data from graphs to answer questions and make decisions or draw conclusions.

A

Tennis matches won this season

Balata Dennery Micoud Roseau

Key

= 4 matches

This graphs shows how four different tennis clubs performed during the season. Which club seems to be the best? Which club won the fewest matches? How many matches did Dennery win?

Number of students taking drawing classes

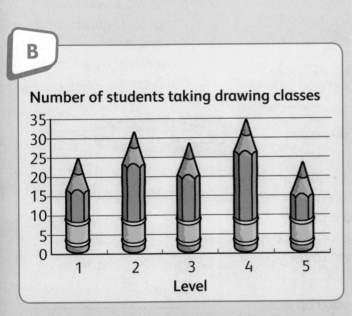

What does this graph show you? Although it has pictures, it is not a pictograph. What type of graph is it? What do the numbers on the vertical scale tell you? How many Level 4s are taking drawing classes?

Think, talk and write

A Tables and graphs
(pages 130–132)

1 Nikki collected data about how much water her friends drink each day.

Micah	$1\frac{1}{2}$ litres
Anne	$\frac{3}{4}$ litre
Zara	2 litres
Tori	$2\frac{1}{4}$ litres
Jess	$1\frac{3}{4}$ litres

2 She decided to draw a pictograph to show the data.

3 What symbol could she use?

4 How would you show each amount using the symbol you've chosen?

B Read and interpret graphs
(pages 133–134)

Work in pairs. Take turns to ask each other questions about this graph.

Answer each other's questions.

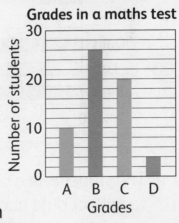

Grades in a maths test

A Tables and graphs

Read through the information to help you remember what you learnt about **pictographs** and **bar graphs** last year.

A pictograph uses small pictures or symbols to show sets of **data**. A pictograph should have a **key** to tell you what each picture or **symbol** represents.

Weekly pizza sales	
Mario's	🍕🍕🍕
Luigi's	🍕🍕🍕
Paolo's	🍕🍕🍕🍕

Key: 🍕 = 100 pizzas

A bar graph uses separate bars to show sets of data. The bars can be **horizontal** or **vertical**. The amount represented by each bar is read from the **scale** of the graph. The bars are labelled to show what set of data each one shows. A bar graph should have a heading to tell you what data it represents.

Bar graphs need a scale on one axis so that you can work out the number of data items (the **frequency**) for each bar.

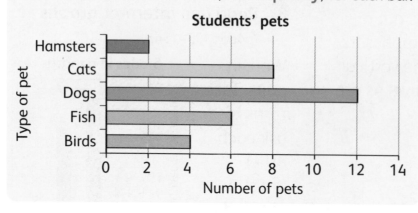

Students' pets

In this unit you will:
* collect data and use tables to organise it
* revise what you know about pictographs and bar graphs
* draw pictographs and bar graphs.

pictographs	vertical
bar graphs	scale
data	frequency
key	collect
symbol	organise
horizontal	

Think back to what you learnt in Topic 8.

Discuss the meanings of these words with your group.

observation	interview
questionnaire	frequency table

1   The students in a class worked on a project to **collect** data about where they live.

   a   How many students live in Half Moon Bay?

   b   How many students were included in the data?

   c   Draw a pictograph to show the data. Choose your own symbol to represent four students.

Community	Students
Grove Village	5
Crescent Drive	2
Half Moon Bay	9
Sugarville	14
Black Bay	9

2   The table shows the marks some students achieved in two tests.

Student	Test 1	Test 2	Total mark
John	72	84	
Anne	52	79	
Marie	53	61	
Gilbert	81	85	

a   Can you tell from the table who got the highest overall mark?

b   Work out the total mark for each student.

c   Draw a bar graph to show your results. Choose a suitable scale for the marks.

3   This table shows the number of pairs of different brands of sneakers that a salesperson sold in one month.

Brand	Pairs sold
Like	28
Buma	23
Padidas	34
Meebok	15
All others	18

a   Decide which type of graph would be most suitable to show this data set and draw it.

b   Explain why you chose that type of graph.

4   Mandy kept track of the weather every day for a full month. She recorded her data on a calendar sheet.

1 Sunny	2 Sunny	3 Sunny	4 Cloudy	5 Rain showers	6 Sunny	7 Windy
8 Sunny	9 Sunny	10 Sunny	11 Sunny	12 Sunny	13 Windy	14 Cloudy
15 Rain showers	16 Rain showers	17 Cloudy	18 Sunny	19 Sunny	20 Rain showers	21 Sunny
22 Sunny	23 Sunny	24 Windy	25 Sunny	26 Sunny	27 Windy	28 Rain showers

a   What month was this? How do you know?

b   Draw up a table to **organise** Mandy's data better.

c   Draw a pictograph to show how many sunny days there were each week.

d   Draw a bar graph to compare the number of windy, cloudy, sunny and rainy days for the month.

131

5   Kelly drew up this table to organise some data she collected by
     interviewing people.

Food	Fish	Chicken	Burgers	Vegetables
Number of people	21	14	12	15

   a   What do you think she was trying to find out? Give a reason for your answer.
   b   How many people did she interview?
   c   Which food was chosen by most people?
   d   Use the data from the table to draw a pictograph. The heading of the
       graph should be: Food we eat most often at home. Choose your own
       symbols and scale.

6   Look at the pizza pictograph on page 138. Redraw this graph as a bar graph.
     Use an appropriate scale and key.

Investigate

7   Choose a topic to research at home or at school.
     You can choose from the ideas in the box or you
     can use an idea of your own.

   a   Decide how you will collect the data you need.
   b   Collect the data and use a suitable table to
       summarise and organise it.
   c   Draw two different graphs to display the data:
       one pictograph and one bar graph. Use a
       different scale for each graph.
   d   Write a few sentences to explain what you
       learnt from your investigation.

Time spent exercising/
doing chores/lazing
around each week

Weather conditions

Method of transport used
to get to school

Favourite song/singer/type
of film/TV programme

Comparison of prices of
different foods/brands
of clothes

What did you learn?

1   Write down the differences between a bar graph and a pictograph.
   a   How do they look different?   b   How do you use them differently?

2   In what ways are a bar graph and a pictograph the same? Discuss your
     answers with your partner.

B Read and interpret graphs

Explain

Do you remember how to **read** and make sense of a graph?

To understand and **interpret** a graph you need to read:
* the **heading** to find out what the graph is showing you
* any **labels** on the graph to find out what the pictures or bars show
* the **key** or **scale** so you can work out what amounts are shown for each set of data.

Key words

read	labels
interpret	key
heading	scale

1 Malia counted the number of butterflies she could see in her garden each day for a week. She drew this graph to display her data.

a On which day did she count the most butterflies? How many did she count?

b On which two days did she count the same number of butterflies? How many did she count on those days?

c How many butterflies did she count altogether on the five days?

d Butterflies like sunlight and wind-free days. On which days do you think this kind of weather was experienced? Why?

Butterflies counted

Monday	✖ 🦋
Tuesday	✖ ✖ ✖ 🦋
Wednesday	✖ ✖ ✖ 🦋
Thursday	✖ ✖ 🦋
Friday	✖ 🦋

✖ = 12 butterflies

2 Read the information on this graph and answer the questions about it.

a How many oranges did vendor A sell?

b Which two vendors sold the same number of oranges? How many did each one sell?

c What was the greatest number of oranges sold by a vendor?

d Which vendor sold less than 30 oranges?

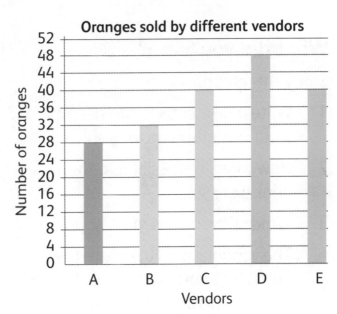

Oranges sold by different vendors

e   These five vendors all sell oranges around the same school. Why do you think some vendors sell more than others? Try to think of at least two reasons.

3   Compare this graph with the one in Question 2 and then answer the questions.

a   How is this graph different from the one in Question 2?

b   Which axis shows the amount represented by each bar?

c   What does the other axis show?

d   How many books did Candy read?

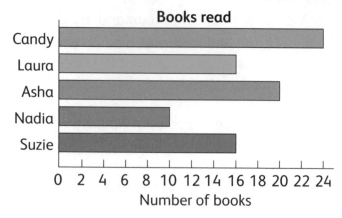

e   How many more books did Candy read than Nadia?

f   Who read more books, Laura or Asha?

4   Asha redrew the graph above using a different scale. This is her graph.

a   How does changing the scale change what the graph looks like?

b   Asha made one mistake on her graph. What was it?

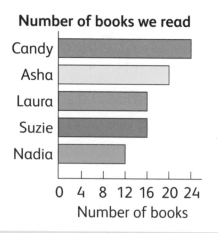

What did you learn?

Study this graph.

1   Write a few sentences to describe what this graph shows you.

2   How many bottles of lemon drink were sold?

3   Which flavour was most popular?

4   The vendor wants to reduce the number of flavours she sells. Which four flavours should she keep? Give a reason for your answer.

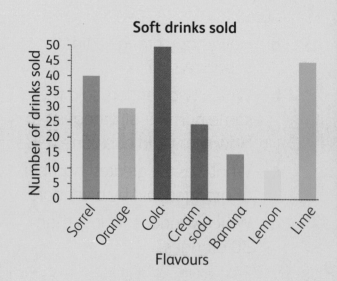

Topic 15 Review

Key ideas and concepts

Make short notes under each of these headings to summarise the main ideas in this topic.

1   Uses of tables   2   Pictographs   3   Bar graphs

4   How to read a graph

Think, talk, write …

1   Which do you find easier to read – pictographs or bar graphs? Why?

2   Is it possible for two bar graphs that represent the same data to look quite different? Explain your answer.

Quick check

1   Study the graph and answer the questions.

Food collected for hurricane relief	
Class 1	
Class 2	
Class 3	
Class 4	

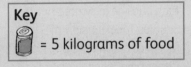

Key

= 5 kilograms of food

a   What does 1 tin mean?

b   How much food did class 4 collect?

c   How would you show $7\frac{1}{2}$ kilograms of food using this symbol?

2   Draw a bar graph to show the data from the graph in Question 1. Choose a suitable scale and label your graph correctly.

A

Teaching notes

Symmetry

* Symmetrical shapes have two halves that are mirror images of each other. If you fold the shapes along a line of symmetry, the two halves will fit onto each other exactly.

Lines of symmetry

* Lines of symmetry are also called mirror lines. If you place a small mirror on the line of symmetry, the reflection in the mirror will show you whether the shape is symmetrical or not.

* Many shapes have more than one line of symmetry. Squares, for example, have four lines of symmetry. Circles have infinitely many. Lines of symmetry can be horizontal, vertical or diagonal.

Drawing symmetrical shapes

* When you are asked to complete a symmetrical diagram, it may be difficult to visualise what the 'other half' of a shape or pattern looks like. It helps to trace the shape and to fold it along its mirror line so you can see what the other half looks like.

* It is also a good idea to use a small mirror to see the orientation of the other half of a symmetrical diagram.

These buildings in Kuala Lumpur in Malaysia are the highest twin towers in the world. They are also a good example of symmetry. What makes them symmetrical?

Justin wanted to know
about lines of symmetry in
rectangles so he did some
experiments with a mirror.
What did he do? What did
his experiments show him?

Think, talk and write

A Lines of symmetry (pages 138–139)

1 Find a rectangular piece of paper.
With a partner, try to fold it in half so
that the two halves fit perfectly on
top of each other in as many ways
as possible.
  a How many ways did you find?
  b Draw a line on the folds that give
     you a perfect match. How many
     lines do you have?

2 Are these letters symmetrical or not?
Give a reason for your answer.
  A   B   C   D   E

B Drawing symmetrical shapes (page 140)

1 The students in a class were asked
to draw the other half of this
symmetrical shape.

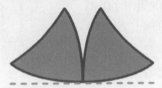

2 Here are four students' drawings.

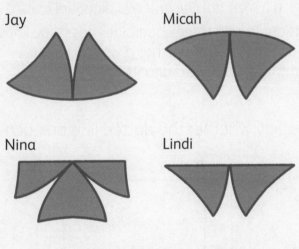

  a Whose drawing is correct?
  b Why are the others not correct?

A Lines of symmetry

A **line of symmetry** is a line that divides a figure into two halves. The two halves are exactly the same, but they face in opposite directions because they are a **mirror image** of each other.

We say a figure is **symmetrical**, or the figure has **symmetry**, if it has a line of symmetry. If you fold the figure in half along the line of symmetry, the two parts match exactly.

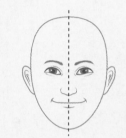

This drawing of a face is symmetrical. The dotted line is the line of symmetry.

Human faces look symmetrical, but they aren't really. There are small differences between the left and right halves of your face. You can check this using a mirror.

A symmetrical figure can have more than one line of symmetry. This letter H has two lines of symmetry, a **vertical** line of symmetry (the blue line) and a **horizontal** line of symmetry (the red line).

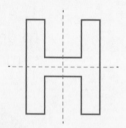

Some figures also have **diagonal** lines of symmetry.

A square has four lines of symmetry: a vertical line, a horizontal line and two diagonal lines.

A triangle with three equal sides has three lines of symmetry. This one has a vertical line of symmetry and two diagonal lines.

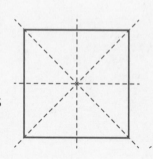
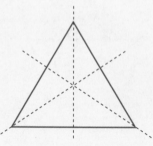

Maths ideas

In this unit you will:
* find lines of symmetry on drawings and other items
* draw lines of symmetry on shapes and letters
* explain what a line of symmetry means.

Key words

line of symmetry
mirror image
symmetrical
symmetry
vertical
horizontal
diagonal

1 Say whether the dotted line on each shape is a line of symmetry or not.

a

b

c

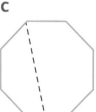

d

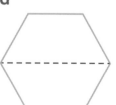

e

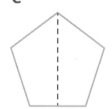

2   The dotted line on the first shape is a line of symmetry. Which diagram (A, B or C) shows the correct mirror image?

a

A   B   C

b

A

B

C

c

A   B   C

Problem-solving

3   The line on each set of letters is a line of symmetry. Can you work out what each word is?

a RED   b   c   d 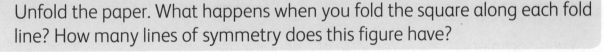   e

BOOK

Practical task

You will need three square pieces of paper and a pencil.

1   Fold the first square in half. Then draw a shape along the folded line, as shown in diagram A.

A

    Cut out the shape and unfold the paper. What do you notice about the shape that you have cut out? How many lines of symmetry does this figure have?

B

2   Now fold the second square in half twice, to get a smaller square, as in diagram B. Draw and cut out any design along the two fold lines.

    Unfold the paper. What happens when you fold the square along each fold line? How many lines of symmetry does this figure have?

What did you learn?

Which of the coloured lines on this shape are lines of symmetry?

139

B Drawing symmetrical shapes

When you are asked to complete a **symmetrical** figure, you have to draw the **mirror image** on the other side of the line (or lines) of symmetry.

Always look carefully at the part of the diagram you are given. If you cannot easily see what the mirror image will look like, you can trace the part you are given. Turn the tracing over to see what you need to draw.

You can also use a small mirror to see what the mirror image looks like. When you place the mirror on the **line of symmetry**, you will be able to see the other half of the figure reflected in the mirror.

Maths ideas

In this unit you will:
* complete drawings of symmetrical figures.

Key words

symmetrical
mirror image
line of symmetry
horizontal
vertical
diagonal

1 Trace these shapes and their line of symmetry. Complete the shapes by drawing the other side.

a 　　b 　　c

2 Draw half of three patterns or shapes for your partner to complete. The line of symmetry on your drawing can be **horizontal**, **vertical** or **diagonal**.

What did you learn?

Can you complete these symmetrical calculations?

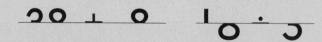

Topic 16 Review

Key ideas and concepts

Copy the diagrams into your book.

Add a heading and labels to each diagram to summarise what you learnt in this topic.

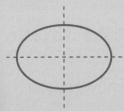 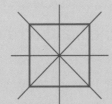 

Think, talk, write ...

1   A square is a quadrilateral (it has four sides) and it has four lines of symmetry. Does this mean that all quadrilaterals have four lines of symmetry? Discuss this in groups. Share your ideas.

2   There are many examples of symmetry in daily life. Find two examples of symmetrical buildings in your community. Sketch them or take photos of them and show your group how they are symmetrical.

Quick check

1   How many lines of symmetry does each of these shapes have?

2   a   Sketch your national flag. Is it symmetrical?

   b   Design and draw two flags. The flags can be any shape. One of the designs should have symmetry, the other should not be symmetrical.

Problem-solving

3   The dotted line on each incomplete shape is a line of symmetry. Work out how many sides the complete shape would have and name each shape.

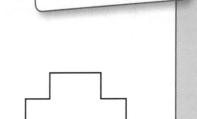

Test yourself (3)

Complete this test to check that you have understood and can manage the work covered this year.
Revise any sections that you find difficult.

1 **a** Write the number nine thousand nine hundred and ninety-nine in a place value table.

    b If you add one to the number, what is your new number? Show this on your place value table.

2 Write these numbers in words.

    a 9 451   **b** 9 001   **c** 4 660

3 Look at the pattern shown here.

    Term 1   Term 2   Term 3

  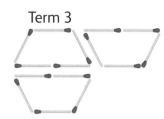

    a What is the rule for the pattern?

    b How many matches will you need to make the 5th term?

    c Which term in the pattern will use 55 matches?

4 Write the equivalent mass.

    a 15 kg = __ g   **b** 4 500 g = __ kg   **c** 500 g = __ kg

5 Round off each number to the nearest thousand.

    a 4 541   **b** 500   **c** 2 834   **d** 1 499   **e** 8 367

6 A fact family has a quotient of 6 and a dividend of 5.

    a What is the other number in the family?

    b Write the fact family for these three numbers.

    c Which number is a product?

7 Look at these shapes.

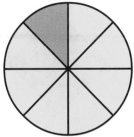 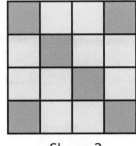 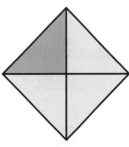

Shape 1   Shape 2   Shape 3

  a What fraction of each shape is blue?

  b Order the fractions from the smallest to the greatest.

  c Write two equivalent fractions for the blue part of each shape.

  d What fraction of shape 2 is yellow? Write this fraction in its simplest form.

8 Josh has entered a 5 km race. He decides to run every day to get fit for the race. On day 1, he manages to run 3 500 m. Each day, he runs 250 m further. How many days until Josh is able to run 5 km?

9 Write the equivalent lengths.

  a 60 cm = ____ mm   **b** 3 m = ____ cm

  c 50 mm = ____ cm   **d** 4 cm = ____ mm

10 A school orders 12 boxes of 145 crayons. Calculate the total number of crayons showing two different methods.

11 Jessica and Anne decide to throw a party for their friends. Jessica brings 15 packets of chips and 8 cupcakes. Anne brings 12 packets of chips and 12 cupcakes. Seven friends come to their party.

  a How many people are at the party?

  b Jessica and Anne make sure that everyone gets the same number of chip packets. How many packets does each person get?

  c The cupcakes are for the friends to take home. How many cupcakes does each friend take home if they all get the same amount? How many cupcakes are left over?

12 Divide. Show your working.

  a $5\overline{)65}$ **b** $9\overline{)108}$ **c** $7\overline{)85}$ **d** $4\overline{)86}$ **e** $5\overline{)153}$

13 How many lines of symmetry does each shape have?

a

b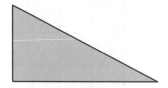

c

d

e

14 The school cricket match starts at 2:00 p.m. and ends at 5:30 p.m. How long is the cricket match?

15 Lauren brings a cake to share with her friends. Anne eats $\frac{1}{2}$ of the cake, Katie eats $\frac{1}{8}$ of the cake and Lauren has $\frac{1}{4}$.

  a How much of the cake was eaten?

  b How much more cake did Anne have than Katie?

  c How much cake is left over?

16 Amy buys three items costing $3.25, $2.00 and $2.75.

  a If she pays with a $10.00 note, how much change will she get?

  b What fraction of her $10.00 did she spend?

  c Can you simplify your answer in part b?

17 Sarah uses 500 g of flour, $\frac{1}{4}$ kg of sugar, $\frac{6}{8}$ kg of bananas and 5 g of baking powder to make her banana muffins.

  a Write the measurements in order from the smallest to the greatest.

  b How much more flour than sugar does Sarah have?

  c What is the total mass of all the ingredients?

18 David wants to put a flower bed in his garden. Right now, he only has a paved backyard. His backyard and a plan of the flower bed are shown below.

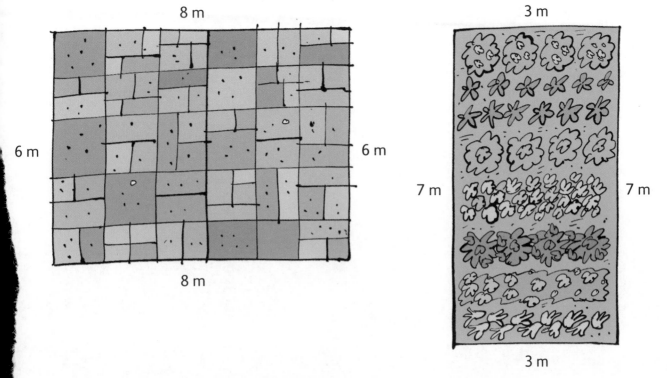

a   What is the perimeter of the backyard?

b   What is the perimeter of the flower bed?

c   Will the flower bed fit into David's backyard? Explain your answer and draw a sketch to show why it will or why it won't fit.

19 How many faces, edges and vertices does a cuboid have?

20 Write the time shown on each clock.

a
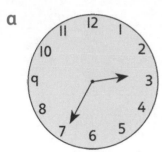

b

21 Mrs Paul bought a television for $3 785 and a lamp for $79.

a   What is the total cost of the two items?

b   How much more was the television than the lamp?